T0236945

Effective Front-End Strategies to Reduce Waste on Construction Projects

Peter G. Rundle · Alireza Bahadori ·
Ken Doust

Effective Front-End Strategies to Reduce Waste on Construction Projects

 Springer

Peter G. Rundle
School of Environment, Science
and Engineering
Southern Cross University
East Lismore, NSW, Australia

Alireza Bahadori
School of Environment, Science
and Engineering
Southern Cross University
East Lismore, NSW, Australia

Ken Doust
School of Environment, Science
and Engineering
Southern Cross University
East Lismore, NSW, Australia

ISBN 978-3-030-12401-4   ISBN 978-3-030-12399-4   (eBook)
https://doi.org/10.1007/978-3-030-12399-4

Library of Congress Control Number: 2018968391

© Springer Nature Switzerland AG 2019
This work is subject to copyright. All rights are reserved by the Publisher, whether the whole or part of the material is concerned, specifically the rights of translation, reprinting, reuse of illustrations, recitation, broadcasting, reproduction on microfilms or in any other physical way, and transmission or information storage and retrieval, electronic adaptation, computer software, or by similar or dissimilar methodology now known or hereafter developed.
The use of general descriptive names, registered names, trademarks, service marks, etc. in this publication does not imply, even in the absence of a specific statement, that such names are exempt from the relevant protective laws and regulations and therefore free for general use.
The publisher, the authors and the editors are safe to assume that the advice and information in this book are believed to be true and accurate at the date of publication. Neither the publisher nor the authors or the editors give a warranty, expressed or implied, with respect to the material contained herein or for any errors or omissions that may have been made. The publisher remains neutral with regard to jurisdictional claims in published maps and institutional affiliations.

This Springer imprint is published by the registered company Springer Nature Switzerland AG
The registered company address is: Gewerbestrasse 11, 6330 Cham, Switzerland

Dedication

Peter G. Rundle wishes to dedicate this book to Mrs. CR, without whose constant empathy and inspiration, the work would have been completed six months earlier.

Preface

This textbook investigates eight engineering construction methodologies that could improve the effectiveness and efficiency of the Australian engineering construction industry. These eight options were: (i) Knowledge Management; (ii) Lean Construction; (iii) Construction Contract Procurement; (iv) Optimal Work Duration on Site; (v) Construction Site Waste; (vi) Rationalization of Construction Safety Regulations; (vii) Sustainable Construction Labour Force; and (viii) Portfolio Project Development. One of these options was selected for detailed research by a peer review panel—"Effective Front-End Strategies to Reduce Waste on Construction Projects".

A global literature review was used to evaluate the relative merits for each of the eight options. A qualitative research method was chosen using a pragmatic framework approach, which employed surveys based on a seminal Australian model (Faniran and Caban 1998) to determine the sources of construction waste. Severity testing was the primary tool used to determine the potential sources of site waste. A survey from Yates (2013) was adapted for examining front-end strategies for reducing construction waste. Thematic analysis was used to evaluate these front-end site waste reduction strategies to synthesize respondent data by developing themes and sub-themes. This resulted in a rich pool of interpreted data.

Design changes were found to be the major source of site waste from Australian construction projects. Early project planning, particularly in the design phase, was considered to be the most likely strategy to result in the front-end minimization of site waste. The top sources of site waste identified in this research were the same as the main potential sources detailed in the global research. The 48 respondents for this research provided no further strategies (other than those previously identified) for front-end site waste minimization.

The "management" theme was the overarching theme for both sources of site waste and front-end waste issues. The critical sub-themes were "quality control shortfalls", for sources of site waste, and "front-end engineering design", for site waste minimization strategies.

The key conclusion of this research was that the Australian governments must look towards the implementation of a suitable regulatory framework that establishes a mandatory code of practice for site waste management.

East Lismore, Australia   Peter G. Rundle
                                                                        Alireza Bahadori
                                                                             Ken Doust

Acknowledgements

Peter G. Rundle wishes to acknowledge his co-authors' assistance: Dr. Bahadori for his unwavering support and Dr. Doust for his lateral thinking. The guidance from Professor W. Boyd, in developing a suitable research framework that ensured that the tail didn't wag the dog, was appreciated. Special thanks to Mrs. Monica Rundle-Hayton for sharing her research experience using thematic analysis, along with SCU faculty academics, Dr. Novak and Deputy Mayor Battista, for their constructive critiques. Sincerest gratitude is expressed to the peer review panel who provided expert option recommendations. Finally, Mr. Rundle thanks the qualitative survey respondents who contributed such thoughtful input.

Contents

Abbreviations

ABC	Australian Broadcasting Corporation
ABS	Australian Bureau of Statistics
ACA	Australian Constructors Association
ACIF	Australian Construction Industry Forum
AGDE	Australian Government Department of Employment
AIG	Australian Industry Group
BCITO	Building and Construction Industry Organization
BIM	Building Information Modelling
C&D	Construction and Demolition
CBA	Cost–Benefit Analysis
CBD	Central Business District
CEO	Chief Executive Officer
CIB	Conseil International du Bâtiment
CIOB	Chartered Institute of Building
DA	Decision Analysis
DIIS	Australian Government Department of Industry, Innovation and Science
ECI	Early Contractor Involvement
EMS	Engineering Management Standard
ENR	Engineering News-Record
EPCM	Engineering, Procurement and Construction Management
EPC	Engineering, Procurement and Construction
EU	European Union
FEED	Front-end Engineering Design
FY	Financial Year
GDP	Gross Domestic Product
GFC	Global Financial Crisis
HSE	Health, Safety and Environmental
HV	High Voltage
ICT	Information and Communications Technology

ILO	International Labour Organization
ISCA	Infrastructure Sustainability Council of Australia
ISO	International Organization for Standardization
JIT	"Just in time"
KM	Knowledge Management
KPI	Key Performance Indicator
KT DA	Kepner Tregoe Decision Analysis
LC	Lean Construction
MBA	Master Builders Association
MBAV	Master Builders Association of Victoria
MC	Main Contractor
NPD	New Project Development
NRUHSE	National Research University—Higher School of Economics
OECD	Organisation for Economic Co-operation and Development
OH&S	Occupational Health and Safety
PFI	Public Finance Initiative
PM	Project Management
PMC	Project Management Consultant
PPM	Project Portfolio Management
PPP	Public–Private Partnership
QMS	Quality Management System
R&D	Research and Development
SCM	Supply Chain Management
SCU	Southern Cross University
SDG	Sustainability Development Goals
SOE	*"State of the Environment Overview"*
TSS	Temporary Skilled Shortage Visa
UK	United Kingdom
UNSW	University of New South Wales
USA	United States of America
WMP	Waste Minimization Plan

List of Figures

List of Tables

List of Charts

About the Authors

Peter G. Rundle is Chartered Professional Civil Engineer. He has held roles as engineering manager, construction manager, project manager, project director and company director, in the development and execution of major engineering projects, on 12 long-term assignments in 10 countries. He holds an MBA, MSc. (Research) and a DCE. He is undertaking a PhD in 2019 to develop this research. He was a 2017 finalist for *SCU Alumnus of the Year* and earned a High Distinction average for his MBA, topping all students in three postgraduate units. In 2016, he attended London School of Economics in the UK and studied a summer school unit of Macroeconomics. He has been engaged on a broad range of projects, varying from a site manager on a US$2 billion dollar nickel mine construction project, to engineering manager for a port project in the South Pacific. He spent the early part of his career as a bridge engineer, working in such locations as Papua New Guinea and Saudi Arabia. He has recently been Chief Operating Officer for the largest division of a global engineering company with over 2000 personnel reporting to him. He has been a study manager, project manager and project director on 22 project studies. The last feasibility study he supervised was for a US$4.5 billion coal mine with a study cost US $58 million. He is Member of both Engineers Australia and also the American Society of Engineers. He has been employed by several of the world's largest contractors (Bechtel; Brown and Root; Aker Kvaerner, etc.) both in Australia and internationally.

Alireza Bahadori PhD, CEng, MIChemE, CPEng, MIEAust, RPEQ, NER is Research Staff Member in the School of Environment, Science and Engineering at Southern Cross University, East Lismore, NSW, Australia. He received his PhD from Curtin University, Perth, Western Australia. During the past 20 years, he has held various process and petroleum engineering positions and was involved in many large-scale projects. His multiple books have been published by multiple major publishers, including Springer, Elsevier and Wiley. He is Chartered Engineer (CEng) and Chartered Member of Institution of Chemical Engineers, London, UK (MIChemE); Chartered Professional Engineer (CPEng) and Chartered Member of Institution of Engineers, Australia; Registered Professional Engineer of Queensland (RPEQ); Registered Chartered Engineer of Engineering Council of UK and Engineers Australia's National Engineering Register (NER).

Ken Doust is Director of Windana Research Pty Ltd; Co-Director of UCCRN Australian Oceania Hub; Co-Ordinator of Eng. Management Program at Southern Cross University. His research areas include:

- Planning designing and enabling sustainability and climate change appropriate infrastructure;
- Asset management practices and infrastructure planning and maintenance;
- High-speed train systems;
- Railway track geometry condition management;
- Long-term corporate knowledge of State Rail, RIC and RailCorp, spanning 30 years, with previous management and analyst roles within engineering, regions and planning divisions;
- Strategic planning roles within City Rail and in private sector. Experience includes both conceptual and process thinking.

Chapter 1
Introduction

1.1 Context

The Australian engineering construction industry is the largest non-services sector in this economy, accounting for 8.1 per cent of the gross domestic product (GDP) (Gruszka 2017, p. 4). However, during 2014–15, the Australian Bureau of Statistics (ABS) ranked the "rate of innovation" in the Australian construction sector as the third lowest of all businesses (Australian Bureau of Statistics (ABS) 2015). Competitive advantage, gained by adopting effective and efficient engineering construction methodologies, is important within the Australian engineering sector. This is because many international engineering consultants and contractors enter the local market due to our stable political, social and economic foundations (Austrade 2018; Gruszka 2017). There is a relatively low entry barrier for the Australian engineering construction sector, which accounts for the ready access to the local market available for foreign firms (IBISWorld 2017). Therefore, Australian engineering construction companies must continue to adapt and innovate to maintain a competitive advantage through effective and efficient operability (Thompson et al. 2014).

The purpose of this research has been a staged process that entailed the initial identification of eight methodologies that had the potential to enhance effectiveness and efficiency within the Australian engineering construction industry. These eight options were identified after an initial literature review and the selection criteria were based on six key performance indicators (KPIs) derived for this study. A key criterion of option identification was a requirement that all identified options must provide value in the context of a triple bottom line philosophy (Slaper and Hall 2011), which dictates that tangible commercial benefits to key stakeholders, as well as ecological and social benefits, can be enjoyed with the implementation of these proposed methodologies.

These eight options, forming the "long-list", were further researched via a global literature review, to collect a sufficient level of information to be used as assessment briefing documents. This assessment entailed an independent review by nine eminent engineering construction specialists, who formed a peer review panel to determine a

© Springer Nature Switzerland AG 2019
P. G. Rundle et al., *Effective Front-End Strategies to Reduce Waste on Construction Projects*, https://doi.org/10.1007/978-3-030-12399-4_1

"short-list" of four methodologies, which would be further researched in academic, practitioner, trade and government literature.

A risk-based decision analysis process was adopted, and scorecards were transmitted via the Internet to the peer review panel, which comprised a broad spectrum of engineering construction specialists from the construction infrastructure and commercial building, engineering design, architectural design, finance, commercial, legal, major client, environmental and planning sectors of the engineering construction industry. This decision analysis process was overviewed by a Southern Cross University (SCU) engineering school senior lecturer. The "short-list" evaluation by the author, Peter G. Rundle (the author), was included with the independent peer review "short-list" scorecard. After reviewing the peer review panel consolidated scorecards, the author provided written summation as to why the consolidated decision analysis was followed for all but one of the "short-list" options. The author then prepared a closing analysis as to why the four rejected options were unsuitable for further review, in accordance with the research KPI parameters.

Finally, the two Australian engineering construction peak industry bodies, the Australian Construction Industry Forum and a cohort of engineering construction specialists were consulted about the appropriateness of the four "short-listed" options, regarding industry appetite for further development. This completed the first phase of the research.

The second stage of the works comprised of a more extensive literature review with an emphasis on investigating such matters as the voids in the academic literature (particularly in Australia) and the most appropriate compliance with KPIs, to determine a "Go Forward" case for detailed research. The "Go Forward" case was the option selected for further investigation. A rigorous Kepner and Tregoe (2013) risk-based decision analysis process was adopted for use by the second independent peer review panel (consisting of a combination of engineer academics, heavy infrastructure construction and commercial building practitioners and government representation). A Skype teleconference workshop was convened and this peer review panel collaborated with an independent workshop facilitator to choose a "Go Forward" case using the Kepner Tregoe decision analysis (KT DA) scorecard system. The author had the right to alter the peer review panel recommendations. The author independently scored the same "Go Forward" case and readily concurred with the peer review panel's advice. The author attended the workshop as technical advisor to address panel queries and his assessment was not integrated into the panel scorecard.

The "Go Forward" case was entitled "*Front-End Strategies to Reduce Waste on Australian Construction Projects*". For the purposes of this textbook, front-end strategies are those strategies that entail early involvement in the preliminary project design and development phase (BHP Billiton 2007). The author then embarked on an additional review of the literature to validate that the "Go Forward" case, independently selected by the KT DA, was indeed the most appropriate option to take into the final detailed research phase. Further analysis was undertaken of the three rejected options via a review of the literature to address peer review panel queries and develop the possible research scope for the front-end site construction waste "Go Forward" case.

The third and final phase was the detailed research for the "Go Forward" option, in order to further fill the void in the academic literature in the Australian field on possible solutions for the reduction of construction site waste at the front-end of the supply chain (rather than the off-site supply chain back-end options, involving landfill, re-use and recycling) During this final literature review stage, a preliminary research framework was also constructed. Most discussions were held with senior engineering construction executives from the Australian Construction Industry Forum (ACIF) and Australian Constructors Association (ACA) peak industry groups, as well as a select cohort of construction specialists, to confirm that there was an appetite in the local engineering construction industry to build on the knowledge acquisition data surrounding front-end construction waste minimization. This check to gauge the industry's market interest was carried out to ensure that the large gap in the local academic literature on this option was not due to a lack of interest in the topic.

The research framework for this textbook was developed for final use under the valued guidance of a Southern Cross University (SCU) professor, as recommended during the SCU confirmation panel process. After an extensive literature review of research methods, including quantitative, qualitative and mix method approaches, the author selected a qualitative pragmatic framework approach, using thematic analysis to synthesize survey participant responses. Respondent data would be obtained via a survey questionnaire and the authors decided that previously published peer-reviewed research surveys would be the appropriate path forward. The three-part survey polled data on: (i) respondent demographics; (ii) the sources of construction site waste; and (iii) front-end site waste minimization strategies.

The survey sample size for this qualitative research was suggested as between 10 (Riemen 1986) and 30 (Creswell 2013), with 54 completed surveys eventually being provided by respondents from a sample pool of 102 participants. Survey questionnaires were adapted for use from Faniran and Caban (1998), the Australian seminal academic paper on sources of construction waste, and the USA Yates (2013) paper, which considered possible front-end construction waste mitigation strategies and was funded by a peak contracting industry body in the USA.

During the SCU confirmation process, it was recommended that the author carry out an assessment into the degree of difficulty expected for the execution of a cost–benefit analysis (CBA) on front-end site waste reduction (which was proposed to form part of the detailed research phase) and present this brief to an assigned SCU economics professor for guidance. A review of the limited academic literature available on construction waste CBA research was undertaken during November 2017. The proposed CBA would evaluate the economic, ecological and environmental benefits that could occur in Australia due to the front-end reduction of construction site waste.

The proposed CBA research scope was tabled in discussions with the assigned economics professor, who advised that a work of such magnitude and complexity would be beyond the scope of this textbook. However, it could be considered as a suitable topic for high-level Ph.D. research in future. All three authors concurred with the professor's valued written objective assessment. Accordingly, the CBA was excised from the scope of the research.

In conclusion, the body of work that comprises this textbook should be seen in the context of a three-stage research process that has evaluated eight, and then four possible strategies that could enhance the effective and efficient execution of Australian engineering construction projects and ultimately concludes with detailed research of the "Go Forward" case. Pursuant to this approach, at confirmation stage, textbook was renamed *"Effective Front-End Strategies to Reduce Waste on Australian Construction Projects"*. The reader's attention is drawn in Fig. 1.1, "Textbook Research Schematic", which provides a visual presentation of the research workflow.

1.2 Definition of Effectiveness and Efficiency

As a precursor to investigating the literature to determine the eight methodologies to be evaluated in this research, it was necessary to define "efficiency" and "effectiveness" in a construction engineering sense, with respect to the book title. Accordingly, the preliminary academic literature review was devoted to finding an appropriate description of these two concepts. A detailed body of work has been executed by Chau (1993), Langston and Best (2001) and Langston (2014) on defining "efficiency". Each paper has built on the existing body of knowledge to progressively develop a mathematical model that defines construction "efficiency". Chau (1993) based his model on project performance rate rather than industry productivity. Langston and Best (2001) developed a "performance index", based on output (production capacity) to input (resource consumption), that attempts to combine time and cost into one performance indicator via an upgraded mathematical model. Finally, Langston (2014) later computed construction "efficiency" via a simple equation (construction efficiency = production capacity × resource consumption) to undertake a landmark study comparing USA and Australian construction efficiencies for one of the largest project samples ever, using high-rise buildings, constructed of 20 storeys or more, as the sample comprising 150 skyscrapers in Australia and 354 skyscrapers in the USA. However, the use of a specifically "hard" quantitative commercial building mathematical model is deemed too narrow a definition for the purposes of this broad study.

Theoretical concepts for "effectiveness" were considered in a literature review. Wysocki (2001) argues in his book, *"Effective Project Management"* that various tools can be utilized to enhance "effective" project implementation on a single project basis, without defining what "effectiveness" was. "Efficiency" was not considered in this book; however, Wysocki (2001) notes that improved project "effectiveness" will result in improved schedule and budget outcomes. Hyvari (2006) refers to project management (PM) "effectiveness" as being related to organizational structure and project manager personality characteristics. Sundqvist et al. (2014) argue that the optimal method to evaluate project-based "effectiveness" and "efficiency" is to use these two factors as a lens, to view the project improvement process via a focus on

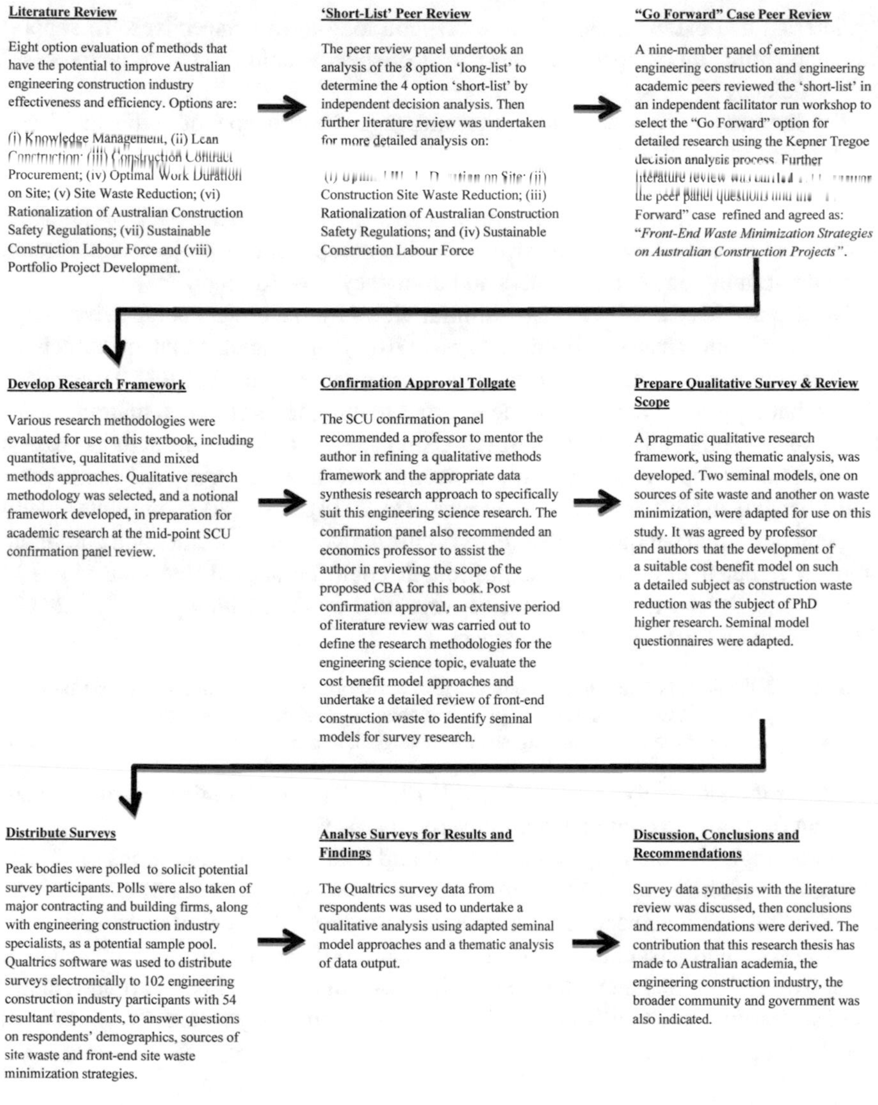

Literature Review

Eight option evaluation of methods that have the potential to improve Australian engineering construction industry effectiveness and efficiency. Options are:

(i) Knowledge Management, (ii) Lean Construction; (iii) Construction Contract Procurement; (iv) Optimal Work Duration on Site; (v) Site Waste Reduction; (vi) Rationalization of Australian Construction Safety Regulations; (vii) Sustainable Construction Labour Force and (viii) Portfolio Project Development.

'Short-List' Peer Review

The peer review panel undertook an analysis of the 8 option 'long-list' to determine the 4 option 'short-list' by independent decision analysis. Then further literature review was undertaken for more detailed analysis on:

(i) Optimal Work Duration on Site; (ii) Construction Site Waste Reduction; (iii) Rationalization of Australian Construction Safety Regulations; and (iv) Sustainable Construction Labour Force

"Go Forward" Case Peer Review

A nine-member panel of eminent engineering construction and engineering academic peers reviewed the 'short-list' in an independent facilitator run workshop to select the "Go Forward" option for detailed research using the Kepner Tregoe decision analysis process. Further literature review was carried out to resolve the peer panel questions that the "Go Forward" case refined and agreed as: *"Front-End Waste Minimization Strategies on Australian Construction Projects".*

Develop Research Framework

Various research methodologies were evaluated for use on this textbook, including quantitative, qualitative and mixed methods approaches. Qualitative research methodology was selected, and a notional framework developed, in preparation for academic research at the mid-point SCU confirmation panel review.

Confirmation Approval Tollgate

The SCU confirmation panel recommended a professor to mentor the author in refining a qualitative methods framework and the appropriate data synthesis research approach to specifically suit this engineering science research. The confirmation panel also recommended an economics professor to assist the author in reviewing the scope of the proposed CBA for this book. Post confirmation approval, an extensive period of literature review was carried out to define the research methodologies for the engineering science topic, evaluate the cost benefit model approaches and undertake a detailed review of front-end construction waste to identify seminal models for survey research.

Prepare Qualitative Survey & Review Scope

A pragmatic qualitative research framework, using thematic analysis, was developed. Two seminal models, one on sources of site waste and another on waste minimization, were adapted for use on this study. It was agreed by professor and authors that the development of a suitable cost benefit model on such a detailed subject as construction waste reduction was the subject of PhD higher research. Seminal model questionnaires were adapted.

Distribute Surveys

Peak bodies were polled to solicit potential survey participants. Polls were also taken of major contracting and building firms, along with engineering construction industry specialists, as a potential sample pool. Qualtrics software was used to distribute surveys electronically to 102 engineering construction industry participants with 54 resultant respondents, to answer questions on respondents' demographics, sources of site waste and front-end site waste minimization strategies.

Analyse Surveys for Results and Findings

The Qualtrics survey data from respondents was used to undertake a qualitative analysis using adapted seminal model approaches and a thematic analysis of data output.

Discussion, Conclusions and Recommendations

Survey data synthesis with the literature review was discussed, then conclusions and recommendations were derived. The contribution that this research thesis has made to Australian academia, the engineering construction industry, the broader community and government was also indicated.

Fig. 1.1 Textbook research schematic

the internal and external, as well as short- and long-term perspectives. To support this supposition, their paper considers a project-specific definition of "effectiveness" and "efficiency".

The literature review has shown that although the concepts of "efficiency" and "effectiveness" are often used, they are rarely defined. Indeed, the two terms are often applied interchangeably (Sundqvist et al. 2014). Certain literature applies these terms when describing competencies for project execution (Lampel 2001), while other researchers describe how to improve a specific component of project management using the terminology "effectiveness and efficiency" (Ward 1999).

DeToro and McCabe (1997) have similar views from a project perspective on the definition of "effectiveness" (as does Parast (2011)) and indicate that an "effective" product satisfies or exceeds external customer requirements and that "efficiency" shows that a product meets all of the internal requirements for cost, margins, asset utilization and other efficiency measures. Using these definitions, the authors suggest that continuous improvement of projects within portfolios and for general comparison would occur using this approach (Sundqvist et al. 2014). Rose (2005) presents several different definitions of quality that focus on the delivery of a specification compliant product, rather than customer satisfaction. It could be argued that both views of "effectiveness" lead to the same result—the final product (Sundqvist et al. 2014). Sundqvist et al. (2014, p. 280) state that:

> from an academic and practitioner point of view, clarifying what is meant when discussing efficiency and effectiveness related to project management should be the starting point for discussion, measuring and evaluating project management processes.

The paper aims to apply "effective" and "efficient" methodologies in project management by applying continuous improvement processes.

Accordingly, for the purposes of precise and unambiguous definitions of "effectiveness" and "efficiency" in this book, "effectiveness" shall be defined as: "satisfying or exceeding customer needs with a compliant product"; this definition suitably synergizes both Sundqvist et al. (2014) and Rose's (2005) complementary concepts of "effectiveness". "Efficiency" can be described as: "meeting all internal requirements for cost, margins, asset utilization and other related efficiency measures" (Sundqvist et al. 2014, p. 281).

1.3   Initial Research Question

The next step in the research, after defining the topic title, was to develop the preliminary research question, which would be valid during the first two phases of the work, up to commencement of the detailed research phase.

Although formal approval to proceed with this research was granted on the 23 March 2017, the author commenced, part-time, on a preliminary literature review from the 14 February 2017 and was consequently able to prepare this initial research question shortly after notification to proceed on the 23 March 2017. At research commencement, the study title was "*Effective and Efficient Methodologies in the Australian Engineering Construction Industry*". A methodology from Swinburne University (2016) was adapted to determine the initial research question; refer to Table 1.1.

As noted elsewhere in this chapter, "Introduction", the triple bottom line philosophy is a key selection criterion for choosing appropriate methodologies to enhance construction industry effectiveness and efficiency. Accordingly, the initial research question is defined as: "*what effective and efficient methodologies are available to add triple bottom line value to the Australian engineering construction industry?*".

1.4   Key Performance Indicators (KPIs)

After the definition of the initial research question, key performance indicators (KPIs) were developed for use as metrics to assist in the objective selection of the eight methodologies that formed the eight options "long-list". There were other criteria that these options had to comply with, namely a demonstrated void in academic literature (particularly in Australia), and meeting the definition requirements of what constitutes a methodology that has the potential to improve engineering construction industry effectiveness and efficiency, as described in Sect. 1.2, "Definition of Effectiveness and Efficiency".

The six KPIs for this research project were carefully evolved, so as to be of maximum benefit to the industry, the ecology and the community at large, to be readily implemented and to benefit parties across a broad spectrum. These KPIs are noted as follows: (i) the textbook will be of transparent commercial benefit to contractor/engineering houses, client end-users and the community at large; (ii) the proffered solutions can be readily and therefore, expediently, implemented; (iii) the topics selected for investigation shall provide maximum stakeholder benefits; (iv) the solutions to the identified inefficiencies and ineffective practices are, by and large, available within the academic and professional international body of knowledge;

Table 1.1   Initial research question construct

Initial research question construct		
Research question criterion	Criterion description	Remarks
Interrogative word	What....?	The interrogative "what", explicitly pre-supposes a process
Phenomenon	Effective and efficient methodologies	–
Context	Australian engineering construction industry	–

By P. Rundle, 2017 (adapted from: Swinburne 2016)

(v) the research must be practical in nature and address a void in the Australian engineering construction business and the work must be valuable to this industry; and finally, (vi) the identified research must broadly comply with a triple bottom line philosophy (Slaper and Hall 2011) and that commercial, social and ecological benefits will be provided by these options.

References

ABS. (2015). *Annual report, 2014–2015*, cat. 1000.1, ABS, Canberra.

Austrade. (2018). *Building & construction capability report*, viewed 6 September 2018, https://www.austrade.gov.au/International/Buy/Australian-industry-capabilities/Building-and-Construction.

Billiton, B. H. P. (2007). *Project manual for capital investment projects*. Australia: BHP Billiton Project Management Centre.

Chau, K. W. (1993). Estimating industry-level productivity trends in the building industry from building cost and price data. *Construction Management and Economics, 11*(5), 370–383.

Creswell, J. W. (Ed.). (2013). *Qualitative inquiry and research design—choosing among five approaches*. USA: Sage Publications.

DeToro, L., & McCabe, T. (1997). How to stay flexible and elude fads. *Quality Progress, 40*(8), 55–60.

Faniran, O. O., & Caban, G. (1998). Minimizing waste on project construction sites. *Engineering Construction and Architectural Management, 17*(1), 57–72.

Gruszka, A. (2017). How technology is transforming Australia's construction sector. *StartupAUS Report*, pp. 1–67, viewed 6 September 2018, http://www.apcc.gov.au/ALLAPCC/SAUS_ConTech_Report_2017.pdf.

Hyvari, I. (2006). Project management effectiveness in project-oriented business organizations. *International Journal of Project Management, 24*(3), 216–225.

IBISWorld. (2017). *Australia industry report (ANZSIC)—engineering consulting services*, viewed 5 September 2018, www.clients.IBISworld.com.au.exproxy.scu.edu.au.

Kepner, C. H., & Tregoe, B. B. (2013). *The new rational manager*. Publishing, Princeton, USA: Kepner Tregoe Inc.

Lampel, J. (2001). The core competencies of effective project execution: The challenge of diversity. *International Journal of Project Management, 19*(8), 471–483.

Langston, C., & Best, R. (2001). An investigation into the construction performance of high-rise commercial office buildings worldwide based on productivity and resource consumption. *International Journal of Construction Management, 1*(1), 57–76.

Langston, C. (2014). Construction efficiency: A tale of two countries. *Energy, Construction & Architectural Management, 21*(3), 320–325.

Parast, M. (2011). The effect of six sigma projects on innovation and firm performance. *International Journal of Project Management, 29*(1), 45–55.

Riemen, D. (1986). The essential structure of a caring interaction: Doing phenomenology. In P. Munhall & C. Oiler (Eds.), *Nursing research: A qualitative perspective* (pp. 85–105). USA: Appleton-Century-Crofts.

Rose, K. H. (2005). *Project quality management: Why, what and how*. USA: Ross Publishing.

Slaper, T., & Hall, T. (2011). The triple bottom line: What is it and how does it work. *Indiana Business Review, 8*(1), 4–8.

Sundqvist, E., Backlund, F., & Chroneer, D. (2014). What is project efficiency and effectiveness? *Procedia—Social and Behavioral Sciences, 119*, 278–287.

Swinburne University. (2016). *How to write a research question*. SU School of Electrical Engineering, viewed March to October 2017, https://www.youtube.com/watch?v=lJS03FZj4K.

Thompson, A., Peteraf, M., Gamble, E., & Strickland, A. J., III. (2014). *Crafting & executing strategy*. New York: McGraw Hill/Irwin.

Ward, S. (1999). Requirements for an effective project risk management process. *Project Management Journal, 30*(3), 37–43.

Wysocki, R. (Ed.). (2001). *Effective project management: Traditional, agile and extreme*. USA: Wiley

Tates, J. (2013). Sustainable methods for waste minimisation in construction. *Construction Innovation, 13*(3), 281–301.

Chapter 2
"Long-List" of Eight Methodologies

2.1 Option Selection Background

The preliminary literature review focused on locating suitable construction methodologies that could be of benefit to the Australian engineering construction industry. Based on their collective engineering experience, and after a preliminary confirming assessment of the academic literature, the three topics that the authors believed could add value to the industry were: (i) Optimal Work Duration on Site; (ii) Construction Site Waste Reduction; and (iii) Sustainable Construction Labour Force in Australia.

The authors determined that eight engineering construction methodologies would form the option "long-list", which would be independently evaluated down to a four-option "short-list" after a detailed literature review; these four methodologies would then form the basis of a "short-list". The next step of the literature review involved finding the five other most suitable options to make up the eight-option "long-list". In order to evaluate the literature to locate these five innovative options, it was deemed necessary to determine what "innovation" meant, particularly in the context of this research related to engineering construction. This was because it was suggested in certain literature that innovative project processes were deemed to provide improvements in efficiency and effective project performance (Gann 2000; Slaughter 1993). Similarly, to other terms presently in vogue (such as "sustainability", "effectiveness" and "efficiency") "innovation" can be construed as having an imprecise meaning.

A review of the literature indicated that there has been a degree of academic innovation in the field of construction engineering, particularly in the Organisation for Economic Co-operation and Development (OECD) countries (Suprun and Stewart 2015). Innovation usually leads to reduced construction costs and/or project schedule improvements (National Research University—Higher School of Economics (NRUHSE) 2013). Gann (2000) suggests that higher levels of construction innovation shall improve productivity, while Tatum (1991) posits that profitability will increase as a result of employing innovative construction methods. The international construction sector has been defined as a "laggard industry" in terms of innovative development due to the inherent high-risk nature of the work, along with the low

© Springer Nature Switzerland AG 2019
P. G. Rundle et al., *Effective Front-End Strategies to Reduce Waste on Construction Projects*, https://doi.org/10.1007/978-3-030-12399-4_2

margin nature of the work (Dibner and Lerner 1992; Slaughter 1993). Several studies have denoted the importance of innovation in improving construction outcomes and yet have highlighted the apparent disconnect with a failure to diffuse these new innovations into the construction industry (Davidson 2013; Egbu 2004; Sayfullina 2010).

National spending on innovation is not great, even in OECD countries. As an example, Australia spends only 2.4 per cent of its gross domestic product (GDP) on Research and Development (R&D), while Germany's R&D spends is 2.8 per cent of GDP, and the USA spends 2.9 per cent on its R&D (Hampson et al. 2014). Coupled with the inherent risk-adverse nature of the construction business, firms are conservative in their approach to innovative solutions (Miller et al. 2009). Accordingly, the diffusion of all available knowledge on efficient and effective construction methodologies available to the construction world, rather than a massive expansion in innovative construction engineering research and development via a dramatic increase in expenditure, would prove an altogether more effective and efficient means of addressing this situation (Slaughter 1993). Accordingly, this textbook will endeavour to identify the most appropriate, available, innovative, effective and efficient processes through the review of academic literature.

This existing body of knowledge of the selected streams shall be objectively determined for further research and diffused into the construction sector. Means of expanding construction innovation, such as commercial incentives, taxation privileges and low interest loans, is therefore not in the scope of this research. The literature suggests that there are three key reasons for the lack of innovation diffusion in the Australian construction industry (Australian Government Productivity Committee 2007; Bruneela et al. 2010; Hall et al. 2001; Suprun and Stewart 2015). These reasons are: (a) government-owned institutions generally pay lower remuneration than multinational innovation companies, so there is a large staff turnover in government R&D service; (b) R&D diffusion relies on the personal contribution of one author who turns a good idea into a viable proposition—this person may not have the necessary marketing skills; and (c) governments usually fund R&D at a local level, rather than advanced innovation such as construction and Information Communications Technology (ICT).

Myers and Marquis (1969) suggest that "innovation" is the initial commercial transaction of a new technological process. General definitions of innovation can also be applied to construction innovation, namely, innovation:

I. Is something new (Freeman 1974; Schumpeter 1947).
II. Can implement "change" (Abernathy and Clark 1985).
III. Is the first construction firm to use the particular "new" technology (Tatum 1987).
IV. Shall derive benefits to other stakeholders involved in a construction project (Ling 2003).
V. Possibly creates a higher potential risk (Dodgson 2000).

Slaughter (1993) classified construction innovation typology as incremental, radical, architectural, modular and system. Murphy et al. (2008, p. 102) state that "Slaughter's (1998) examples of these innovations were drawn primarily from civil engineering projects. The innovations contributed to the buildability and functionality of a building, e.g. on-site construction processes". Blayse and Manley (2004) note that the Slaughter (1993) definition of construction innovation is the generally accepted description. Slaughter (1993, p. 83) defines innovation as follows:

> Innovation is the actual use of a nontrivial change and improvement in a process, product, or system that is novel to the institution developing the change. Innovation in the construction industry can take many forms.

Slaughter (1993) characterizes such innovation according to whether it is "incremental" (small, and based on existing experience and knowledge), "radical" (a breakthrough in science or technology), "modular" (a change in concept within a component only), "architectural" (a change in links to other components or systems) or "system" (multiple, integrated innovations) (Blayse and Manley 2004, p. 144).

The Slaughter model (1993) was utilized when considering the five other methodologies from an innovation perspective. After a broad search of academic, practitioner, trade and government literature, five other innovative methodologies were selected for review as the balance of the eight options forming the "long-list". Another criterion used for preliminary assessment purposes in forming the "long-list" was probable compliance with the triple bottom line strategy of the options adding commercial, social and ecological value, post-implementation. The other five methodologies that were added to the list, in an Australian context, were: (i) Knowledge Management; (ii) Lean Construction; (iii) Construction Contract Procurement Practices; (iv) Rationalization of Australian Construction Safety Regulations; and (v) Portfolio Project Development. Next, KPIs were prepared as parameters, with which the eight projects would be evaluated. Refer to Sect. 2.2, "Eight Option 'Long-List'", for these details.

These eight selected streams, chosen after completion of the preliminary literature review, form the "long-list" of options, from which a "short-list" of four options for further research was objectively chosen using an independent peer review panel, adopting a risk-based framework for selection. This risk-based decision analysis framework was also utilized in the selection of the "Go Forward" option(s) for detailed research that formed the basis of this book (Kepner and Tregoe 2013). The "long-list" of items that have the potential to enhance Australian construction engineering industry efficiency and effectiveness is: (i) Knowledge Management; (ii) Lean Construction; (iii) Construction Contract Procurement Practices; (iv) Optimal Work Duration on Site; (v) Construction Site Waste; (vi) Rationalization of Australian Construction Safety Regulations; (vii) Sustainable Construction Labour Force; and (viii) Portfolio Project Development. These eight options were appraised in the context of an Australian environment.

These eight options were double-checked, preliminarily, against the six KPIs in Sect. 1.4, "Key Performance Indicators (KPIs)", to verify suitability for the further literature review. At the completion of the literature review, the KPIs were used again to check the viability of these eight "long-list" options, moving into the "short-list" option phase. Another critical tollgate criterion was the requirement that the "Go

Forward" option(s) for further detailed research clearly demonstrated a void in the academic literature. These eight options were researched in the literature to prepare sufficient information and primarily to provide a briefing paper to the independent peer review panel who would undertake a risk-based analysis to decide which four options of these eight should form the "short-list". The initial literature review of these eight options, prepared as a briefing document for use by the independent peer review panel, in "short-list" determination, is detailed in Sect. 2.2, "Eight-Option "Long-List"".

2.2 Eight-Option "Long-List"

This section provides the initial literature review of eight methodologies (interchangeably known as "options" or "cases"), which form a "long-list" of eight engineering construction methodologies that will potentially improve the effective and efficient operability of the Australian engineering construction industry. This knowledge acquisition database, of the eight options constructed by the author progressive literature review, will be used as a briefing paper by an independent peer review panel of construction specialists to objectively select and make recommendations for a four-option "short-list" for a more detailed literature review.

2.2.1 Knowledge Management

Proper preparation of a project, prior to execution, is crucial for construction companies as a method of identifying potential risks and solutions. Project preparation processes can be greatly improved using the management of knowledge by engineering construction specialists (Ribeiro and Ferreira 2010). Teece et al. (1997) argue that improving project performance in a market of uncertain demand is critical to business survival in a rapidly changing economic and financial environment. Further, firms that learn to capture, organize, combine and share their traditional resources and capabilities, in new and distinctive ways, provide value for their customers and, in general, their shareholders, compared to their competitors (Ribeiro and Ferreira 2010; Teece et al. 1997). Knowledge Management (KM) enhances this organizational competitiveness. Companies need to create a corporate culture that allows their employees' knowledge to be converted into organizational knowledge, enabling informed decision-making through knowledge flow (Zhuge 2006).

However, in practice, transferring knowledge within the construction sector is historically difficult to achieve (Argote et al. 2000). Construction projects generate a huge body of knowledge available for sharing, which can be re-used within the organization, across all projects, and can improve a company's overall performance, simply by employees sharing best practices, lessons learned, experiences, insights and new knowledge (Kogut and Zander 1993). Notwithstanding this, despite the vast amount of published research findings, the facts plainly indicate that KM projects rarely meet their project objectives and are often abandoned prior to completion (King et al. 2008; Wing and Chua 2005). Accordingly, a methodology that involves the use of empirical modelling, in which the primary and primitive ingredient is experience, could be developed (Beynon et al. 2002). As an example, Yin (2003) posits that a qualitative case study approach would provide a body of knowledge from various participants sharing knowledge that could be identified and analysed.

According to Owen and Burstein (2005), the organizational culture of an engineering consultant and/or contracting firm will play a critical role in developing a KM system that encourages employees to become members of a formal community of practice, rather than an informal personal network for knowledge distribution. They also maintain that personnel training will need to be employed as the KM system is developed within an organization, to ensure system maintenance and utilization, as well as the integration of tacit knowledge and existing networks in the KM system. Ferrada and Serpell (2013) argue that their research, carried out on three case studies, conclusively shows that the selection of the appropriate construction methods to delivery projects is primarily a result of personal experience. A KM system approach would allow a company to make more informed and objective construction method choices that would benefit clients by ensuring project objectives were met (effectiveness), as well as optimizing project contractor cost and schedule milestones (efficiency) (Ferrada and Serpell 2013; Halpin 2006).

2.2.2 Lean Construction

Koskela and Howell (2002) posit that Lean Construction is an adaption of manufacturing production techniques for use on construction projects. Lean Construction (LC) takes a holistic approach to a project, which includes initial project development, detailed design, construction implementation, commissioning, operational life, salvage and recycling (Abdelhamid et al. 2008). The LC approach attempts to manage construction processes for minimum cost (efficiency) and provide maximum customer value (effectiveness). Ballard and Howell (1994) observed that, typically, only 50 per cent of activities on a project's weekly schedule were being achieved, and that classic construction management models were failing to achieve project objectives. LC has the following priorities:

I. Keep work flowing and crews productive.
II. Reduce inventories of tools and materials.
III. Reduce costs (Sowards 2004).

LC endeavours to engage client, designer, engineering consultant, subconsultants, contractors and sub-contractors, along with other stakeholders, early on in the project, in order to align project objectives. Additionally, another unique feature of the LC system is the use of an integrated "last planner" schedule that includes engineering design consultants, clients, contractors and suppliers, along with other interested parties' activities, on the one programme (Mossman 2013). Two fundamental differences between LC and traditional project management (PM) practices are that: (i) LC processes continually strive to improve both product quality and project work flow, while traditional PM does not, and (ii) LC processes are developed so that contract variations are minimized, but classic PM processes do not try and mitigate scope changes (Koskela and Howell 2002).

Nowotarskia et al. (2016) note that another feature of the LC process is the relentless drive to reduce material wastage, which is also a feature of the Toyota Production manufacturing process to provide the client with exactly what they require, and not what Toyota thinks the client requires. In their research, a commercial high-rise project in Poland identified three high commercial risk processes during concept development that had the potential to result in budget blowouts (Nowotarskia et al. 2016). The project decided to strategically adopt LC procedures for these three activities which comprised of: (i) column concrete pours; (ii) scaffold ordering; and (ii) the site laydown area. The Nowotarskia et al.'s (2016) analysis, post-project completion, showed that the LC processes adopted on the three identified potential high-risk activities resulted in significant savings over traditional construction management processes; although in some instances, accurate cost variances were noted as readily comparable.

The use of LC in construction organizations has been limited so far. However, a growing number of construction companies are in the process of implementing quality assurance and total quality control measures. Notwithstanding this, LC still lags behind the manufacturing industry in performance results. The barriers preventing LC in progressing include a relative lack of competition in construction amongst international contractors, and a slow response by academic institutions in recognizing LC's capabilities to deliver projects more effectively and efficiently. These barriers would be only temporary if the philosophy of LC eventually proves even more successful (Alarcon 1997).

2.2.3 Construction Contract Procurement Practices

Naoum and Egbu (2016) argue that a construction project is successful if schedule, budget estimate and quality criteria are all fulfilled and, increasingly, contract methods of construction delivery can shape project success. Construction professionals are looking for alternative contractual agreements for construction procurement, as external drivers (such as inflation and recession) and internal drivers (such as increasingly more complex projects and client expectations with higher-quality hurdles yet demand for lower cost outcomes); both cause problems and create new pressures on successful project delivery (Naoum and Egbu 2016, p. 310).

There are several new procurement mechanisms that have come into use [such as "Public Private Partnership" (PPP), "Alliance Partnering" and "Public Finance Initiative" (PFI)] to address the current international construction climate. Lupton et al. (2007) describe these agreements as an approach to procurement, while Naoum

(2003) defines them as a philosophy, built on a contract of trust. By and large, traditional forms of contracts (e.g. lump sum and schedule of rates), with cost as the key tender selection criterion, continue to dominate construction procurement agreement documentation (Adekunle et al. 2009). During the worldwide resource commodities boom between 2001 and 2014, Australia was the eighth largest engineering design consulting market in the world in 2012, with 8.2 per cent expenditure of total international design revenue (Gross 2012).

Australia is home to the world's sixth largest design consultant, Worley Parsons [Engineering News Record (ENR 2016)]. Australia's major contractor, CIMIC (formerly, Leighton Holdings), is the globe's fourteenth largest heavy constructor (ENR 2017). In the past 15 years, 10 of the world's top 30 engineering design consultants have acquired large Australian engineering consulting firms (Australian Constructors Association (ACA) 2017).

Australia has been at the forefront of implementing PPP contracts (Suprun and Stewart 2015) as well as Alliance Partnering, with the world's largest contractor, Bechtel, constructing several major process and mining facilities, as a means of controlling project outcomes via the provision of tangible incentives to all parties: clients, Bechtel, contractors, field workers and stakeholder employees on occasion total up to 3000 personnel (Suprun and Stewart 2015). Current external drivers of change (political, economic, social, technological, environmental and legal forces) impacting upon construction developments create new forms of procurement contractual arrangements in response to these drivers of change. Love et al. (2008), in an academic paper studying public sector procurement processes in Queensland and Western Australia on state government projects, argue that a traditional lump sum may stifle technological innovation, particularly, the design and constructability of public sector buildings.

The Australian and USA construction scenes embrace constructability techniques to save costs, reduce potential safety hazards and accelerate schedules, along with providing the client with an optimal product based on meeting client project business objectives (Naoum and Egbu 2016). Doloi (2013) highlights that this constructability approach integrates all of the key stakeholders across the project life cycle, including trade sub-contractors and sub-consultants. The other modern issue that a corresponding modern approach to contract procurement needs to focus upon is supply chain management (SCM). Love et al. (2004), Isatto et al. (2013) and Azambuja et al. (2014), all believe that SCM must be considered a fundamental organizational component of a project strategy and that a seamless supply chain should be developed during the early project phase. However, it is recognized that because of the fragmented nature of project construction, there are barriers to supply chain integration that overarching procurement plans need to consider (Dainty et al. 2001).

Pressure from clients to improve quality, reduce cost and construct more rapidly has driven the construction industry to seek solutions through innovation (Naoum and Egbu 2016). As early as the middle 1960s, Bowley (1966) recognized the requirement to innovate in order to stay ahead of competitors in the built environment theatre. Abbot et al. (2006) argue that innovation should not just be limited to technical construction aspects, but also to actual construction organizational structure.

Value engineering is a systematic approach to collaboratively evaluating the whole project life cycle by clients, contractors, designers and major vendors, over a series of workshops, coordinated by an experienced facilitator, to optimize the project (Egan 1998). This is done by breaking up the usual fragmented approach to project development and consolidating the process into an integrated method of procurement, communication and idea formulation (i.e. brainstorming).

There are several other key issues, such as sustainability, which were uncovered in the literature review and are now impacting upon Construction Contract choices, and which would be evaluated if this option was selected for further detailed research. However, despite whether the prime contract is:

I.   A traditional fragmented agreement comprising separate contractual arrangements for each party including designers, contractors, sub-contractors, vendors and statutory authorities;
II. A fully integrated design and construct/turn-key arrangement;
III. A partially integrated management/EPCM (engineering, procurement and construction management) contract; or
IV. A partnering alliance arrangement.

Modern pressures require the development of corresponding new and more appropriate procurement contracts.

In addition, clients are looking towards more sophisticated tender selection requirements to respond to a more demanding and knowledgeable public at large. Issues such as sustainability and social justice considerations (i.e. the triple bottom line) also weigh upon client decision-making.

2.2.4 Optimal Hours of Work on Australian Construction Sites

The preliminary literature review has shown that limited research has been undertaken on construction labour force shift duration and days of work on site. Usually, when projects are running late, site hours are increased to between 10 and 12 h per day, and at times, crews are expected to work seven days per week (Intergraph 2012). Intergraph (2012, p. 5) states that:

> [s]cheduling heavy overtime for extended periods can not only double or triple the cost of the work [in a] standard 40-hour work week, but actually produce less completed work due to productivity loss. Owners could easily pay for over 100 hours of work and get less than 40 hours of productive work. Extended work weeks drop productivity after only a few weeks and continue to diminish rapidly after a few months.

They go on to cite an example where an extended work week over a 16-week period resulted in productivity losses of 65 per cent.

The 40-h work week is a construct of the British socialist movement during the industrial revolution. The Stone Mason Guild was provided the eight-hour day in the

1850s in Victoria, Australia, and the push for an eight-hour work day can be traced back to the British movement (Love 2006). No academic papers have been found that consider physiological research into whether the 40-h week is actually the optimum work duration, or simply a heuristic benchmark. However, the evidence shows that there is a correlation between working a 12-h shift on site and a propensity for an increase in incidents and accidents (Lemna et al. 2007).

The Australian construction industry employs just over one million employees and represents seven per cent of GDP (ABS 2008; Australian Government Department of Employment (AGDE) 2015). Historically, the gender-biased nature of construction, as being men's work (Byrne et al. 2005), coupled with a rigid adherence to long hours and inflexible work schedules, has hindered the ability to attract and retain talented staff (Townsend et al. 2011). The Australian construction industry's attempts to move to a five-day working week are seen as "radical", and OECD figures indicate that Australian full-time employees work some of the longest hours of any industrialized country (Campbell 2005). The ABS (2007) reports that 30 per cent of employees work overtime each week, of which 43 per cent are unpaid for this overtime. This situation can be compared to a construction site where manual workers are recompensed for overtime, while engineers and managers (staff employees) are not. There is a mismatch between management and workers, as construction sites move from a six down to a five-day work week, per shift cycle. The mismatch occurs in a five-day week cycle because manual workers have less paid overtime, and construction site management does not lose overtime payment on weekends as it is not in contract, but gains a leisure day, resulting in improved work/life balance. Case studies (Townsend et al. 2011) indicate that there are few, if any, negative inputs as a result of this five-day-per-week initiative.

Clearly, there is further research to be done on optimizing labour shifts and days of work on construction sites. There was a landmark study carried out by Yi and Chan (2014) of Hong Kong construction workers, by which the researchers developed a Monte Carlo simulation-based algorithm to optimize work/rest schedule in the hot and humid Hong Kong environment and ran 21 work patterns to develop the ideal shift under these conditions. Further work needs to be undertaken on optimization modelling in this area.

2.2.5 Australian Construction Site Waste

A 2011 status report on the management of construction and demolition waste in Australia ("Construction and Development Waste Status Report") was prepared for the Australian Government Department of Sustainability, Environment, Water, Population and Communities and the Queensland Government Department of Environment and Resource Management by Hyder Consulting (Hyder) and its project partners—Encycle Consulting and Sustainable Resource Solutions (Hyder Consulting 2011a). The report addresses the generation, recovery, markets and products for construction and demolition (C&D) waste materials across all eight jurisdictions (states and territories) in Australia.

The Hyder report (2011a), using financial year 2008–09 data, established that 19 million tonnes of C&D waste was generated in Australia (Hyder Consulting 2011a). Of that total, 8.5 million tonnes of C&D waste were disposed of in landfill and an additional 10.5 million tonnes of waste stream was recovered and recycled. This average recovery rate of 55 per cent was quite variable across the states and territories, with the best recovery rate being 75 per cent. Recovery rates were consistent with landfill costs; the higher the landfill fee, the higher the recovery rate. The Hyder report (2011a, n.p.) states that:

> High C&D landfill disposal charges provide strong incentive for high volume and regular generators of C&D waste to source separate materials and allow for easier reprocessing. High landfill disposal levy costs provide an incentive to process mixed C&D waste in order to recover certain high value and high volume components and avoid landfill disposal costs.

The Hyder report (2011a) goes on to highlight several other recommendations in the report that include such initiatives as: artificially adjusting the price of "fresh" construction material stone aggregates to level the price playing field and make recycled aggregates more competitively attractive; a continuation of a roll-out throughout all states and territories, of what amounts to a high fee impost on C&D landfill disposal; and a suggestion that governments structure public works tenders to favour contractors who use recycled materials.

A report commissioned for the Australian Government by Edge Environment (2012) clearly demonstrates that the state government initiatives to reduce construction waste landfill are working. The report shows that in 2004–05 financial year (FY), 7.5 million tonnes of C&D residual waste was disposed in landfill and 7.6 million tonnes of C&D waste was recycled, compared to the previously noted FY 2008–09 figures of 10.5 million tonnes recycled and 8.5 million tonnes disposed in landfill. However, higher landfill site disposal fee charges and the major cost of recycling waste materials, generated by Australian federal and state government sustainable waste initiatives, are added cost burdens on construction contractors and residential/commercial builders and are passed directly onto clients as part of the contract agreement sum.

In addition, the significant capital and operating costs of running recovery and recycling facilities at landfill disposal sites are added cost pressures on a shrinking government taxation base, post-2014 resources commodities boom collapse. There has been a welcome international drive to embrace sustainable processes for both the present and, particularly, the future greater good. A Hyder report (Hyder Consulting 2011b, p. 39) commissioned by the Australian Commonwealth Government, that looked at all forms of waste in Australia, clearly demonstrated the explosion of waste product creation in Australia from FY 2002–03, where there was a total of 32.4 million tonnes of waste produced, to FY 2008–09, with a total national production of 60.6 million tonnes of waste. There has been a relative reduction in construction waste to landfill disposal in Australia over the past decade, at a significant financial cost to the public at large. C&D waste represented 38 per cent of all waste produced in FY 2005–06 (Edge Environment 2012) but dropped to 31 per cent in FY 2008–09 (Hyder Consultants 2011a, b).

However, some literature reports that there are strong opinions on the politicization of waste management by governments in Australia. Hickey (2015, n.p.) states that:

> our governments … often rely on surveys and statistics from unreliable (levy-funded) con-
> sultancy sources who quite often quote each other's misinformation on the actual recycling
> figures for construction waste generation and processing at each state level. … [Waste is]
> difficult to measure, as the private sector traditionally handles the majority of Australia's
> waste from the largest waste generation streams of construction and demolition waste, as
> well as the commercial and industrial waste from our manufacturing sectors.

Ritchie (2016) posits that Australian federal, state and territory governments put out waste management strategies, but then do not bother to create accompanying favourable economic and policy conditions to allow businesses and local governments to achieve these targets. He sums up this situation by noting that "[g]overnments should have the courage of their convictions and put in place the mechanisms to permit the private sector and local governments to achieve those recycling targets" (Ritchie 2016, n.p.). Hickey (2015) contends that some government agencies release waste management policies on their preferred methods of construction recycling technology, without regard to statistics, to support their government's political position on waste management and have been known to dismiss their own consultant's strategic waste infrastructure reports. "This is particularly prevalent when these findings and research doesn't (sic) suit their own governments' political message to the wider and generally uninformed members of the public" (Hickey 2015, n.p.). Pursuant to the above, it can be seen that Australian governments need to de-politicize site waste management and provide as accurate statistics as possible from which to make informed decisions.

Figure 2.1 was prepared by Ritchie (2017) from a data analysis of the Commonwealth Government's 2016 "State of the Environment Overview" report, tabled to parliament in March 2017, in Canberra. This table raises obvious questions regarding the efficacy of this report related to the use of five-year-old data in some instances and having three different years for data set records. The previous "State of the Environment Overview" was tabled by the Commonwealth, five years prior, in 2011.

There is a corresponding wealth of trade, government and academic literature on the sustainable treatment of waste. The extensive review of the available literature has shown that there is extremely limited recent academic literature surrounding strategies that address front-end construction site waste reduction. That is, to reduce the front-end creation of construction waste on site, rather than addressing a downstream volume via recycling or via imposing statutory "brakes", such as cost imposts and price equalization schemes. Zero Waste Scotland (2017) concludes that 13 per cent of construction materials are wasted on site and that one per cent of project costs could be saved by reducing this figure to an acceptable level. Coupled with this, the reduction in proliferate waste of construction materials would also reduce emissions, save transportation costs, and reduce energy and water consumption.

State	Year	Landfilled	Recycled	% Diversion	Years Old	Source
NT	2010/11	280,000	20,000	7%	5	Source: National Waste Report 2013
NSW	2012/13	6,300,000	10,500,000	63%	3	Source: State of the Environment Overview 2015
TAS	2010/11	410,000	240,000	37%	5	Source: National Waste Report 2013
QLD	2014/15	4,765,854	3,673,189	44%	1	Source: State of Waste and Recycling in Queensland Report 2014-15
SA	2013/14	914,000	3,588,000	80%	2	Source: Zero Waste SA – SA 2013-14 Recycling Activity Report
WA	2014/15	3,613,310	2,621,540	42%	1	Source: Recycling Activity in Western Australian 2014-15
VIC	2014/15	4,125,479	8,409,714	67%	1	Source: Victorian Recycling Industry Annual Report 2014-15
ACT	2010/11	200,000	730,000	78%	5	Source: National Waste Report 2013
Total		20,608,643	29,782,443	59%		

Adapted from: Ritchie 2017.

Fig. 2.1 Summary of data used in the commonwealth government's 2016 "*State of the Environment Overview*" report

2.2.6 Rationalization of Australian Construction Safety Regulations

Australia ranks 128 out of 148 countries in the World Economic Forum's international index for regulatory burden. The significant overlap and duplication of regulation across levels of government impose even further costs; it has been estimated that recent reforms targeting 17 areas of overlap and duplication will yield business cost reductions of $4 billion a year (Australia Business Council 2013, p. 6). The Office of the Chief Economist (2015) report on the impact of regulations on Australian business states that there are typically over 60 licenses required to set up a construction business in Australia.

However, the highly regulated occupational health and safety area with different legislation between the Commonwealth, the six states and two territories is an area of particular importance to the construction industry, which is a high risk business from a safety perspective (Safe Work Australia 2012). The Commonwealth Government taskforce on regulation tabled a report in 2006, entitled "*Rethinking Regulation*", in which there was a recommendation to harmonize occupational health and safety (OH&S) regulations across all nine governments, as a result of a strong representation from Australian business, including the construction industry, to achieve this goal. Australia is a world leader in OH&S (Raheem and Hinze 2014). The purpose of research into OH&S regulations pertaining to construction would be predicated on the proviso that there would be no increase in the safety risk profile to workers. A 2014 occupational health and safety report undertaken by RMIT University and commissioned by the Australian Constructors Association stresses the importance of appropriate organizational culture in the construction industry to foster safety. The report notes that:

> [a]ccording to James Reason, (2000), the cultural drivers for [OH&S] become increasingly significant as health and safety performance improvements 'plateau' following the establishment of safety 'hardware' and 'software' (that is, technologies and systems). (Lingard et al. 2014, p. 29)

Indeed, Reason (2000) suggests a key component of a good safety culture is "an abiding concern for failure" (Lingard et al. 2014, p. 11). Companies with a good safety culture are alert and responsive to signals of hazard. Government control by stringent regulatory control can only take a "zero harm" philosophy so far, and ultimately, a culture of safety must be introduced into the construction industry, businesses, work crews and to individual construction workers (Lingard et al. 2014, p. 29). Elements of recent regulation (such as mandatory scaffolding in some states, of all buildings over one storey, preparatory to painting external facades) could be integrated and costs apportioned to these new methodologies versus a review of safety accidents/incidents for this work.

In conclusion, it is perceived that there is a real need to determine whether the current OH&S regulatory regime in Australia, across nine different governments, provides an efficient safety net for the Australian construction industry and its employees, and whether there is a requirement to rationalize these government laws. As a minimum, legislation should be consolidated into a unified national OH&S code for a more effective approach to ultimately save lives and improve project outcomes. There is a possible void in the literature, identified during this preliminary review, which needs to be further researched regarding the antecedents of a Rationalization of Australian Construction Safety Regulations.

2.2.7 Sustainable Australian Construction Labour Force

A sustainable Australian construction labour force is defined as a long-term construction workforce that can robustly handle the country's often disruptive drivers of change. These drivers of change include economic cycles impacted by global market forces, such as the global financial crisis (GFC), Brexit and the USA dollar exchange rate, along with impacts on Australia's export market via global commodity prices (OECD 2017). Other external forces include social changes, such as the demographic increase in Sydney's western suburbs population, which is fuelling an infrastructure boom in that region (Cook 2018).

The Australian building, plumbing and construction industry accounts for approximately 10 per cent of Australia's GDP and employs nearly 10 per cent of Australia's workforce (Australian Government Department of Industry, Innovation and Science (DIIS) 2017). The Australian construction industry accounts for half of the total investment in the provision of investment goods in Australia through construction, including building works, and the industry has problems coping with the inconsistent labour demand caused by economic fluctuation (Toner 2006). Unisearch (2002) states that the shortage of skilled on-site labour in Australia is brought about by a combination of factors, including a reduction of union influence, particularly during the conservative Howard Government period, and the resultant rise of trade subcontractors in the field. This has increased the demand for skilled tradespeople while intensifying competition for tender success, to the detriment of quality, innovation and training imperatives.

A further negative impact on the construction industry has occurred through the federal, state and territory governments' lack of emphasis on trades training during this period of high demand for tradespeople (Unisearch 2002). The June 2017 "Construction Outlook", published by the Australian Constructors Association, pointed to an overall growth of 4.5 per cent in construction workers for the second half of the calendar year 2017 with 2018, further increasing for that year by 1.2 per cent, after an overall reduction in construction labour during 2016 of 4.8 per cent, with labour fluctuations picked up by sub-contract skilled tradesmen (ACA 2017). Annual construction turnover in 2017 was expected to show a robust increase of 4.3 per cent, with a 6.8 per cent increase forecast for calendar year 2018, after the eight per cent downturn in construction in a flat 2016 calendar year. The increase in work since June 2017 onwards is primarily due to the extra public infrastructure expenditure of Australia's eastern states (ACA 2017). It is clear, even from this three-year cycle, that a coherent construction labour force plan needs to be established by the nation to cope with the constant economic fluctuations for an industry with such a substantial impact on GDP growth.

Toner (2006) also discusses a peculiar pre-disposition for tradespeople to work in site positions with a lower qualification requirement than a role they could perform. An explanation for this could be that the difference in rates of pay for tradespeople and trades assistants is so similar that many tradespeople are reluctant to take on the added pressure of a trades person's responsibilities, let alone a supervisory leading hand or foreman's set of duties. This leads to a further lack of skilled personnel available to fill job vacancies. However, further empirical research is needed to verify this view.

Welch (2009) observed that construction workers are highly vulnerable to physical and mental illness. Work Health Victoria (2013) carried out a health assessment of some 176,183 male construction workers in that state, which revealed health issues such as high blood pressure of 33 per cent and high cholesterol levels of 76 per cent. Lifestyle problems encountered included risky alcohol intake by 63 per cent of the construction workforce and smoking by 29 per cent (Lingard et al. 2014). Other studies have found that construction workers have high family conflict levels (Lingard and Francis 2004) and suffer from poor sleep quality (Williams et al. 2006). Australian construction workers are in urgent need of programs to improve their physical and mental health issues and create a more healthy, resilient and productive labour force (Welch 2009). A healthier workforce would likely result in less sick leave and reduced injury potential. Another health-related issue that needs urgent attention in the Australian construction industry is the lack of life/work balance, particularly amongst the management/engineering cohort in the field controlling a project who are subjected to "burn out" (Lingard et al. 2014).

Over the past decade, Australia has experienced a labour shortage, a skill gap, and a growth in the ageing population; the construction sector has not been immune. A recent study in New Zealand in 2013 suggested that older workers can add considerable value to the construction industry because they have the experience and deep knowledge of the trade area that can improve productivity on site. It has been argued in the academic literature that the knowledge retention of an organization is important for the sustainable development and longevity of the organization (Wood et al. 2012). Despite this, a recent paper by Kossen (2008) highlighted that in times of economic downturn, workers over 45 years of age are seen to be less productive and are viewed by management as a burden, reducing opportunities for younger people, and should therefore exit the workforce. This age discrimination appears to be occurring in Australia, as well as in other developed countries. Anecdotal evidence suggests that because younger employees are more economical to employ, more malleable, and more willing to perform duties outside of their scope of practice, older personnel are more likely to be terminated in an economic downturn.

The review of academic, trade and professional papers indicates that there are possible voids in the literature review that could be researched pertaining to the antecedents of construction workforce sustainability.

2.2.8 Portfolio Project Development

The final "long-list" option that has been identified during the preliminary academic review for evaluation entails an assessment of the portfolio line of strategic management "attack" for developing projects. The Enterprise Portfolio Management Council (2009, p. IX) argues that most companies "decide not to decide" during a capital project approval process that can be evaluated by good collegiate relationships, via the requester's political influence, by turn or by a rhetorical coin toss. Project Portfolio Management (PPM) offers a rational method of objectively assessing a project, which directs commensurate specialist resources to efficiently deliver a project investment, in order to effectively meet an organization's strategic goals through the project quality and key performance objectives. PPM is a centralized system of one or more portfolios of projects that identifies, prioritizes, authorizes, manages and controls projects/programs aligned to corporate strategies (Enterprise Portfolio Management Council 2009). Eisenhardt and Zbaracki (1992) believe that a firm's project portfolio is the embodiment of their strategy. Kester et al. (2013, p. 2) note that "to survive in the long run, firms may need to radically refocus their portfolio resource allocations as technology capabilities and competitor offerings change".

The three features of successful portfolios are:

I.   Strategic alignment—the portfolio make-up reflects the firm's strategic business priorities.
II. Balance—the New Project Development (NPD) portfolio is balanced with respect to the different types of projects and their risk/reward profiles.
III. Maximal portfolio value—an optimal ratio between resource input and value creation (Kester et al. 2013).

However, there may be a downside to companies using the portfolio approach to develop projects due to a lack of adequate resources to execute the works, which is often caused by too many low values, minor projects passing hurdle gate milestones (Cooper et al. 2004). Cooper et al. (2004, p. 1) argue that projects must be done right, but just as importantly, the right projects must be done, which is a prescient statement directly relatable to efficient and effective project construction delivery imperatives. In order to make the necessary difficult decisions to proceed with potential projects, or to terminate further study, management requires sufficient solid information to successfully cull project portfolios, and project selection must be made, in parallel with a resource capacity assessment, in order to judge resource supply against pipeline project demand. A study initiated by the Industrial Research Institute, USA, indicates that the main challenges to the portfolio methodology are: (i) poor prioritization of projects—too many projects passing through hurdles; (ii) lack of resource balancing; (iii) lack of solid information to either continue with, or terminate, a project; and (iv) too many minor/low-revenue projects' proceeding.

Interestingly, Cooper et al. (2004) posit that many medium- and high-value projects, requiring relatively low resources, often do not proceed because high-return and high-resource projects are more favoured by management. Accordingly, should this stream proceed, then solutions to any identified pitfalls would also need to be addressed.

2.3 "Long-List" Decision Analysis to Select "Short-List"

2.3.1 "Long-List" Decision Analysis Framework

Professor Stuart Pugh developed a decision-making model known as the Pugh matrix, used to select an optimal choice from a list of alternatives (Pugh 1991). Key criteria are selected for the decision, and alternatives are then ranked against these criteria (Pugh 1991). The Kepner Tregoe process is a commercial version of the seminal Pugh matrix, which is widely used in the business sector to make risk-based decisions (Kepner Tregoe 2017).

An adapted Kepner and Tregoe (2013) risk-based decision analysis has been used to independently determine a four-option "short-list" from the eight-option "long-list" considered in Sect. 2.2. Prior to preparing the eight briefs via literature review in Sect. 2.2, the author indicated at the commencement meeting that it was his intention to ultimately research two options in full. However, after this "long-list" literature review phase, it became apparent that no more than one option would be practical for this textbook.

The first step in this process was the selection of the independent peer review panel who would provide an objective decision analysis of the eight options, relatively scoring each option to allow selection of the four-option "short-list". This peer review panel comprised of engineering construction specialists from a broad array of fields, ranging from construction law through to project management and environmental science skill sets, to allow a comprehensive overview of the value of each option.

The second step in the "long-list" decision analysis process was the preparation of a risk assessment. The author determined that the eight options would be considered "innovative" methodologies in the Australian engineering construction environment. Accordingly, a literature review on risks to innovation assisted in determining the major potential risks to these eight methodologies.

The next step entailed the formulation of the decision analysis matrix for use as a scorecard that would be circulated amongst nine eminent engineering construction specialists who would form the independent peer review panel and would be given the task of reviewing the eight options and scoring them via the decision analysis (DA) matrix scorecard to select the four-option "short-list".

For probity purposes, a SCU senior lecturer, with risk management facilitation experience, acted as the decision analysis facilitator who oversaw this process and was copied in on internet correspondence between the author and the peer review panel. The completed panellist scorecards were transmitted directly to the SCU facilitator by the author.

Table 2.1 Panel participant list for decision analysis

Independent peer review panel participant list for decision
analysis to select "short-list"

Panel member	Highest academic qualifications	Profession
Ms. T. Boyle	B. Econ., and M. Proc.	Commercial manager
Mr. W. LaForest	B. Com., and M. Applied Finance	Financial manager and company treasurer
Mr. S. Burvill	B. Eng. Sc., and B. Law	Corporate counsel
Mr. V. Kawecki	B. Eng. Sc., and M. Eng. Sc.	Cost, planning and scheduling
Mr. G. Stacey	B. Eng. Sc.	Heavy construction management
Mr. R. Miranda	B. Eng. Sc.	Design project management
Ms. L. Tucker	B. Sc. Hons. Geology	Environmental scientist
Mr. J. Howe	Certificate III Eng.	Contractor general superintendent—Heavy construction
Ms. V. Tamansa	B. Arch.	Project architect

By P. Rundle, June 2017

2.3.2 "Long-List" Peer Review Panel

Details of the independent peer review panel of nine eminent practitioners, from various engineering construction disciplines, are provided in Table 2.1. The author chose the panel participants from a broad spectrum of engineering construction fields. This was because it was the author's sampling strategy to have an array of views from varying perspectives in order to obtain the best four options for further research based on an overarching view from capital work owners, contractual, financial, legal, design and environmental engineering construction specialists, rather than simply a construction site perspective. This approach was adopted to endeavour to give each of the options an opportunity to be viewed on their full merits.

So as to engage the nine-person review panel, these participants were given an opportunity to comment on both the marking criteria applicability and the suitability of marking criterion weighting for the decision analysis scorecard and the risk assessment; several panellists critiqued the literature review and other panellists had constructive suggestions on other viable options. All nine independent panellists provided completed scorecards between 23 June and 3 July 2017. In addition, the author independently scored these options, which have been integrated into the findings. The author has experience in engineering studies, detailed design, construction and management of transportation infrastructure, mining resources and commercial building, on both greenfield and brownfield projects.

The co-authors and the facilitator suggested that the author's scorecard also be combined with the nine independent peer review panel results, and the author concurred. The facilitator collated results were then circulated by the author to peer reviewers and co-authors. Full details of this process follow below.

2.3.3 "Long List" Option Risk Assessment

The author posits that, as the eight methodologies under evaluation would be considered as innovations in the Australian engineering construction industry, it shall be necessary to undertake a further review of academic literature to ascertain the prime potential risks to engineering construction innovation and other capital projects containing elements of innovation. Manley (2006) (based on research by the Australian Cooperative Research Centre for Construction Innovation from 2003 to 2005, which surveyed 400 companies, 14 government bodies, eight industry associations and four universities) stated that the single largest obstacle to innovation in the construction industry is prohibitive cost; this factor is built into the scoring criterion weighting for "long-list" scoring. Scores are five down to one, and effectiveness and efficiency of a particular option are given a high score of five.

Clients (who developed demanding project briefs) and emerging construction site crises/disasters were found to be the two greatest drivers of construction innovation of the above-noted research outcomes (Manley 2006). Blayse and Manley (2004) also include building industry material and equipment manufacturers as key drivers of innovation in the construction sector. Manley (2006) concludes that public sector clients are the major supporters of construction innovation, have the largest budgets for such innovation research and development and are most likely to adopt such innovative construction practices, before private sector client counterparts.

A synthesis of the literature indicates that the three largest risks to innovation are: (i) uncertain demand for the product/services; (ii) innovation capital costs too high; and (iii) financial risk of budget increase too high, particularly with radical innovation (Astebro and Michela 2005; Carbonell and Rodriguez 2006; Voss et al. 2008).

Accordingly, a risk assessment has been undertaken using the following Table 2.2, "Risk Assessment Matrix", to determine a risk profile of the "long-list" of eight options and to assist in the DA for a four-option "short-list" of the three potential highest risks for each of the eight options. A consolidated risk rating of extreme, high, moderate and low was apportioned to each option.

Table 2.3 details the risk assessment undertaken for each of the eight "long-list" items, based on three identified key potential risks related to construction engineering innovation processes. A SCU senior lecturer, with a Ph.D. in the area of risk management, interrogated the risk assessment process detailed herein. Additionally, the nine independent peer review panellists were requested to comment on the risk assessment process. Several panellists are familiar with the risk-based decision analysis process; for example, the construction law specialist is in control of risk management for a 7500-person engineering consultant firm.

Table 2.2 Risk assessment matrix

Likelihood	Impact				
	Insignificant	Minor	Moderate	Major	Severe
Almost certain	Moderate	High	High	Extreme	Extreme
Likely	Moderate	Moderate	High	High	Extreme
Possible	Low	Moderate	Moderate	High	Extreme
Unlikely	Low	Moderate	Moderate	Moderate	High
Rare	Low	Low	Moderate	Moderate	High

Adapted from BHP Billiton Risk (2007)

2.3.4 "Long-List" Decision Analysis Scorecard

The decision analysis (DA) process adapted from the Kepner Tregoe model (2013) has evaluated the marking criteria for all eight options, which synthesizes the KPIs, the research question and the void in the literature requirements. Refer to Table 2.4, which details the marking criteria in the DA pro forma. Not all KPIs were included in these marking criteria, only those vital to the DA process. In any case, all eight options were appraised against all KPIs, prior to this decision analysis tollgate, during the literature review. A score out of five was given to each criterion, five being the most important and a score of one being the least important. These marking criteria scores are constant across each option pro forma and were prepared by the author. During the peer review panel evaluation phase, each panellist was asked to comment on these weightings and the criteria for suitability; these comments were taken on-board on pro forma. Refer to Table 2.4, Column A, for marking criteria weightings.

The peer review panel entered a score of one to five into the "Option Score" in Column B of the DA scorecard for marking criteria items one to seven, inclusive, for all eight-option DA scorecards; five being the highest score and one being the lowest score. Table 2.4 is a "typical" pro forma scorecard that was distributed electronically for all eight "long-list" options to all nine participants. The author opted to provide a score for each of the options for criterion No. 7, "identified void in the literature", as he carried out the literature review and hence, was in the best position to provide a score for this criterion.

Table 2.3 "Long-list" risk assessment

No.	Option	Market demand risk rating	High capital cost risk rating	Budget blowout risk rating	Overall risk rating	Remarks
I	Knowledge management (KM)	H	Y	A	Extreme	Implementation of a KM system will include changing a firm's culture and will incur an inflated cost with potential for budget blowout
II	Lean Construction (LC)	X	X	X	Extreme	Implementation of a LC system will include changing a firm's culture and will incur high costs; market demand is moderate, historically
III	Construction Contract procurement practices	X	X	X	High	Potential is high for budget blowout if innovative contract approach fails during execution. Extra costs will be involved in drafting new approaches. Australia is in the forefront of using new procurement strategies, but demand may be moderate as many new strategies are in vogue
IV	Optimal Work Duration on Site	X	X	X	Low	Low-risk approach with high payback potential
V	Construction Site Waste	X	X	X	Low	Low-risk approach with high payback potential

(continued)

Table 2.3 (continued)

No.	Option	Market demand risk rating	High capital cost risk rating	Budget blowout risk rating	Overall risk rating	Remarks
VI	Rationalization of Australian Construction Safety Regulations	x	x	x	Moderate	There could be resistance from government regulators to rationalize a status quo that would dilute demand appetite
VII	Sustainable Construction Labour Force	x	x	x	Moderate	Considerable public funding is required to promote industry training, and there is entrenched bias about perceived age and gender matters
VIII	Portfolio Project Development	x	x	x	High	Approach is robust, but past implementation has been flawed

By P. Rundle, 2017. Adapted from BHP Billiton (2007), Kepner and Tregoe (2013), Pugh (1991)

The author's score for scorecard item six, "identified void in literature", for each option is noted as follows:

- I.   Knowledge Management (two—no discernible void in literature);
- II.   Lean Construction (two—no discernible void in literature);
- III. Construction Contract Procurement Practices (two—no discernible void in literature);
- IV.   Optimal Work Duration on Site (five—at this stage of literature review, an apparent significant void);
- V.   Construction Site Waste Reduction (five—at this stage of literature review, an apparent significant void);
- VI.   Rationalization of Australian Construction Safety Regulations (three—large quantity of literature but limited on rationalization issue);
- VII. Sustainable Construction Labour Force (three—significant information, however, it is disjointed); and
- VIII. Portfolio Project Development (two—a wealth of practical information, especially from business houses).

Table 2.4 Scorecard for decision analysis of "long-list" eight options to prepare "short-list"

Option No.	Option description	DA scorecard for "short-list" preparation		
No.	Option marking criterion	A: Marking criterion weighting	B: Peer panel member option rating	A × B: Option criterion total
1.	Positive impact of option on efficiency defined as: *meeting all internal requirements for cost, margins, asset utilization and other related efficiency measures* in Australian construction industry	5		
2.	Positive impact of option on effectiveness defined as: *satisfying or exceeding customer needs with a compliant product* in Australian construction industry	5		
3.	Commercial benefit of option to construction industry stakeholders including contractors, suppliers, consultants and customers	4		
4.	Social and environmental benefit of option	3		
5.	Ease of option application and roll-out in the construction industry	3		
6.	Identified void in literature per option (rating by P. Rundle)	5	Pre-filled in by author	
7.	Risk profile of option (rating by P. Rundle)	Pre-filled on each of the eight pro forma reports		
Grand total option no.				

By P. Rundle, June 2017. Adapted from Kepner and Tregoe (2013)

As noted Sect. 2.3.3, ""Long-List" Option Risk Assessment", the risk profile for each option was transposed onto the pro forma for each of the eight options. However, the full suite of risk assessment information was also provided to panellists.

Please note that both Column A and Column B scores out of one to five did not have to be sequential; each marking criterion could have been "five", and each panellists' option score for an option could have been "two" for each criterion.

2.3.5 *"Long-List" Decision Analysis to Determine "Short-List"*

Refer to Appendix C, which summarizes the sum of scores of each of the nine panellists for each of the eight "long-list" options, including the scores of the author. Refer to Table 2.5, which summarizes the eight-option scores.

It is apparent that participants weighed options differently, with some people generally marking with relatively high scores, and other individuals with lower scores. However, in several cases, option rankings were quite similar amongst the panel. Accordingly, it was decided to highlight the top four selections by each participant and consolidate this into a four-option "short-list" as a "sanity check" on simply

Table 2.5 "Short-list" DA scores and ranking

"Short-list" DA scores and option rankings

No.	Option	Grand total each option	Option rank
I	Knowledge management Extreme risk	794	7
II	Lean Construction Extreme risk	701	8
III	Construction Contract Procurement Practices High risk	962	4
IV	Optimal Work Duration on Site Low risk	1107	1
V	Construction Site Waste Low risk	1074	2
VI	Rationalization of Australian Construction Safety Regulations Moderate risk	904	5
VII	Sustainable Construction Labour Force Moderate risk	965	3
VIII	Portfolio Project Development Moderate risk	890	6

By P. Rundle, July 2017

adding the scores. The author evaluated each of the peer review panel and the author own top four rankings individually. The two top options that consistently scored high and have low-risk profiles on the final "short-list" are:

I.   Option 4—Optimal Work Duration on Site (top scores = 10 hits in the top four list); and

II. Option 5—Construction Site Waste (second most scores = eight hits in the top four list).

The next option with the highest score was Option 7, Sustainable Construction Labour, which has the next number of hits (seven hits in the top four list). Option 3, Construction Contract Procurement Practices, with six hits in the top four list was fourth. Option 6, Rationalization of Australian Construction Safety Regulations (five hits in the top four list), came in fifth.

The "sanity check" indicated that both methods of scoring provided exactly that same result in formulating the "short-list". The author decided to delete Option 3, Construction Contract Procurement Practices, from fourth place on the "short-list" and substitute in lieu with Option 6, Rationalization of Australian Construction Safety Regulations, for the following salient reasons:

I.   Although Option 6, Rationalization of Australian Construction Safety Regulations, has been evaluated as a moderate risk option for review, the high risk involved in using new and (legally) untested contractual processes, with high major commercial consequences in the event of failure, must be considered for the contract procurement option.

II. There is a significantly larger void in the literature on safety regulation rationalization on Australian construction sites compared to the literature on innovative engineering construction procurement contracts, which is extensively covered in both Australian academic and professional literature.

III. It is beyond the scope of this book to develop new strategies for procurement contracts. As an example, extensive legal knowledge would be required to undertake this valuable work, which could readily become a PhD dissertation in its own right.

For the record, Option 1, Knowledge Management, scored second lowest overall and received no top four place votes. This option was adjudged as an extreme risk presenting limited opportunities for the discovery of a unique void in the academic literature. *Prima facie*, what has been determined during the literature review on Knowledge Management is that there appears to be a broad disconnect between the high volume of quality academic literature in the construction engineering sector and the lack of an adequate conduit for the dissemination of this knowledge into the industry. Notwithstanding this, there are unique issues in the project development sphere, such as the distinctive footprint of many projects that can make the distribution of knowledge about recycling practices a difficult process.

Additionally, construction has been defined as one of humankind's most complicated fields of endeavour and lends itself to a significant sharing of knowledge through a largely informal communication process of a collegiate and/or anecdotal basis (Ferrada and Serpell 2013; Tully 2016). During the write up on Knowledge Management (KM) earlier in this chapter, it was noted that many KM projects so resoundingly failed that they were abandoned during and at times even prior to implementation. Notwithstanding this, the Australian Constructors Association has recognized the value to the industry of more effective and efficient KM processes and is in the process of commissioning a means for study in this area.

Option 2, Lean Construction, scored lowest of all eight options and never received any top four votes. It was evaluated as having extreme risk potential because of likelihood for cost blowout during implementation, with extremely limited scope for uncovering a void in the literature. It was ascertained through further journal research, post-evaluation in this chapter, that British academic literature in particular has roundly criticized this Japanese production line approach on human capital considerations and related social issues (Green and May 2003). It is argued that the "enterprise culture" business ideology and construction management's "machine mentality" in Lean Construction thinking reduce complex issues to a mechanistic quest for efficiency that shall ultimately sacrifice project quality (Green and May 2003). Green et al. (2008, p. 426) contend that:

> this enterprise culture that came to dominance during the 1980s has had significant (negative) implications for the UK (sic) construction sector with fashionably espoused improvement recipes. Such lean constructions legitimize serving to reinforce the material manifestations of the enterprise culture. In consequence, the UK (sic) industry is characterized by a plethora of hollowed-out firms that have failed to invest in their human capital.

An important feature of this research is the promotion of both sustainability and corporate social responsibility in the development of the two "Go Forward" options, which are issues, it could be argued, to which a Lean Construction approach pays scant regard.

The final stream that was evaluated, Option 7, Portfolio Project Development, received four hits. There is a wealth of academic and trade literature regarding this topic. The issue lies in the adroit execution of this important process, which realistically lies outside the envelope of this textbook. Companies, such as BHP Billiton, have successfully used the portfolio process to effectively implement their enormous capital works pipeline.

In conclusion, the validated four-option "short-list", for further academic literature research, is confirmed as follows:

  I. Optimal Work Duration on Site; Construction Site Waste Reduction;
 II. Rationalization of Australian Construction Safety Regulations; and
III. Sustainable Construction Labour Force.

References

Abbott, C., Jeong, K., & Allen, S. (2006). The economic motivation for innovation small construction companies. *Construction Innovation, 6,* 187–196.

Abdelhamid, T, S., El-Gafy, M., & Salem, O. (2008) Lean construction: Fundamentals and principles. American Professional Constructor Journal 4, 8–19,

Abernathy, W. J., & Clark, K. B. (1985). Innovation mapping: The winds of creative destruction. *Research Policy, 14*(1), 3–22.

ABS. (2008). *Gross Domestic Product,* cat. no. 6202.0. Canberra: ABS.

Adekunle, O., Dickinson, M., Khalfan, M., McDermott, P., & Rowlinson, S. (2009). Construction project procurement routes: an in-depth critique. *International Journal of Managing Projects in Business, 2*(3), 338–354.

Alarcon, L. F. (1997). *Lean construction processes.* Chile: Catholic University Press.

Argote, L., Ingram, P., Levine, J., & Moreland, R. (2000). Knowledge transfer in organizations. *Organizational Behaviour and Human Decision Processes, 82*(1), 1–8.

Astebro, T., & Michela, J. L. (2005). Predictors of the survival of innovations. *Journal of Product Innovation Management, 22*(4), 322–335.

Australian Bureau of Statistics (ABS). (2007). *Employee Overtime,* cat. no. 6342.0. Canberra: ABS.

Australian Constructors Association (ACA). (2017). Australian constructors association construction outlook at June 2017. In *Australian Constructors Association.* http://www.constructors.com.au/wp-content/uploads/2017/08/Construction-Outlook-June-2017.pdf, viewed June–September, 2017.

Australian Government Department of Employment (AGDE). (2015). *May 2015 industry outlook: Construction.* Canberra: Department of Employment, Australian Government.

Australia Business Council. (2013). *Improving Australia's regulatory system.* http://www.bca.com.au/publications/improving-australias-regulatory-system, viewed June, 2017.

Australian Government Productivity Commission. (2007). *Public support for science and innovation—Research report.* Canberra: Productivity Commission.

Azambuja, M., Ponticelli, S., & Brien, W. J. (2014). Strategic procurement practices for the industrial supply chain. *Journal of Construction Engineering Management, 140*(7), 4.

Ballard, G., & Howell, G. (1994). Implementing lean construction: Improving performance behind the shield. In *Proceedings of the 2nd Annual Meeting of the International Group for Lean Construction, Santiago, Chile.*

Beynon, W., Rasmequan, S., & Russ, S. (2002). A new paradigm for computer based decision support. *Decision Support Systems, 22,* 127–142.

BHP Billiton Risk. (2007). *BHP Billiton energy coal hazard and risk procedures–rev 7, BHP Billiton project execution manual.* Australia: BHP Billiton.

Blayse, A., & Manley, K. (2004). Key influences on construction innovation. *Construction Innovation, 4*(3), 143–154.

Bowley, M. (1966). *The British building industry: Four studies in response and resistance to change.* Cambridge: University Press.

Bruneela, J., D'Esteb, P., & Salter, A. (2010). Investigating the factors that diminish the barriers to university–industry collaboration. *Research Policy, 39*(7), 858–868.

Byrne, J., Clark, L., & Van Der Meer, M. (2005). Gender and ethnic minority exclusion from skilled occupations in construction: A Western European comparison. *Construction Management and Economics, 23,* 1025–1034.

Campbell, I. (2005). Long working hours in Australia: Working time regulations and employer pressures. In *Centre for Applied Social Research Working Paper Series,* no. 2005-2. Melbourne: Royal Melbourne Institute of Technology.

Carbonell, P., & Rodriguez, A. I. (2006). The impact of market characteristics and innovation speed on perceptions of positional advantage and new product performance. *International Journal of Research in Marketing, 23*(1), 1–12.

Cook, M. (2018). Danger lurks behind the infrastructure and construction boom. In *Sydney Morning Herald*, July 22. https://www.smh.com.au/business/companies/danger-lurks-behind-the-infrastructure-and-construction-boom-20180718-p4zs8q.html, viewed September 19, 2018.

Cooper, R. G., Edgett, S. J., & Kleinschmidt, E. J. (2004). Benchmarking best NPD practices-II. *Research Technology Management, 47*, 50–59.

Dainty, A. R. J., Briscoe, G. H., & Millett, S. (2001). New perspectives on construction supply management. *Supply Chain Management: An International Journal, 6*(4), 163–173.

Davidson, C. (2013). Innovation in construction—Before the curtain goes up. *Construction Innovation, 13*(4), 344–351.

Dembe, A., Erickson, J. B., Delbos, R. G., & Banks, S. (2005). The impact of overtime and long work hours on occupational injuries and illnesses: New evidence from the United States. *Occupational and Environmental Medicine, 62*(9), 588–597.

Department of Industry, Innovation and Science (DIIS). (2017). *Building/construction fact sheet*. South Australia: DIIS.

Dibner, D. R., & Lerner, A. C. (1992). *Role of public agencies in fostering new technology innovation building*. USA: National Academy of Sciences Press.

Dodgson, M. (2000). *The management of technological innovation: An international and strategic approach*. Oxford, UK: Oxford University Press.

Doloi, H. (2013). Empirical analysis of traditional contracting and relationship agreements for procuring partners in construction projects. *Journal of Management in Engineering, 29*(3), 224–235.

Edge Environment Pty. Ltd. (2012). *Department of sustainability, environment, water, population and communities—Construction and demolition waste guide for recycling and re-use across the supply chain*. https://www.environment.gov.au/system/files/resources/b0ac5ce4-4253-4d2b-b001-0becf84b52b8/files/case-studies.pdf, viewed May–July, 2017.

Egan, J. (1998). *Rethinking construction: Report of the construction task force*. http://constructingexcellence.org.uk/wp-content/uploads/2014/10/rethinking_construction_report.pdf, viewed May–August, 2017. London: HMSO.

Egbu, C. (2004). Managing knowledge and intellectual capital for improved organizational innovations in the construction industry: An examination of critical success factors. *Engineering, Construction and Architectural Management, 11*(5), 301–315.

Eisenhardt, K. M., & Zbaracki, M. J. (1992). Strategic decision-making. *Strategic Management Journal, 13*, 17–37.

ENR. (2016). *ENR top 150 global design and construction firms*. https://www.enr.com/toplists/2016-Top-150-Global-Design-Firms1.asp, viewed January 30, 2018.

ENR. (2017). *ENR top 250 global contractors*. https://www.enr.com/toplists/2017-Top-250-Global-Contractors-1, viewed March, 2017.

Enterprise Management Portfolio Council. (2009). *Project portfolio management—A view from the trenches*. https://leseprobe.buch.de/images-adb/b1/6c/b16c1bb6-190a-45c5-bdef-e37f649d74a6.pdf, viewed April–May, 2017.

Ferrada, X., & Serpell, A. (2013). Using organizational knowledge for the selection of construction methods. *International Journal of Managing Projects in Business, 6*(3), 603–614.

Freeman, C. (1974). *The economics of industrial innovation*. London, UK: Pinter.

Gann, D. M. (2000). *Building innovation: Complex constructs in a changing world*. London, UK: Thomas Telford.

Green, A., & May, S. (2003). Re-engineering construction—Going against the grain. *Building Research and Information Journal, 31*(2), 97–106.

Green, S., Harty, C., Abbas, A. A., Larson, G. D., & Chung, C. K. (2008). On the discourse of construction competitiveness. *Building Research and Information Journal, 36*(5), 426–435.

Gross, A. C. (2012). The global engineering consultancy market. *Business Economics, 47*(4), 285–296.

Hall, B. H., Link, A. N., & Scott, J. T. (2001). Barriers inhibiting industry from partnering with universities: Evidence from the advanced technology program. *Journal of Technology Transfer, 26*(1/2), 87–97.

Halpin, D. (2006). *Construction management*. Hoboken, NJ, USA: Wiley.

Hampson, D., Kraatz, J. A., & Sanchez, A. X. (2014). *R&D investment and impact in the global construction industry*. Oxon: Routledge.

Hinkey, I. (2015). *Landfilling construction waste in Australia*. https://sourceable.net/landfilling-construction-waste-in-australia/, viewed April–December, 2017.

Hyder Consulting. (2011a). *Construction and demolition waste status report management of construction and demolition waste in Australia, Queensland Government*. http://www.environment.gov.au/system/files/resources/323e8f22-1a8a-4245-a09c-006644d3bd51/files/construction-waste.pdf?, viewed May–July, 2017.

Hyder Consulting. (2011b). *Department of sustainability, environment, water, population and communities—Waste and recycling in Australia 2011 incorporating a revised method for compiling waste and recycling data, Australian Commonwealth Government*. http://www.wmaa.asn.au/event-documents/2012skm/swp/m2/1.Waste-and-Recycling-in-Australia-Hyder-2011.pdf, viewed May–July, 2017.

Intergraph. (2012). *Factors affecting construction labor productivity managing efficiency in work planning white paper*. https://www.intergraph.com/assets/global/documents/SPC_LaborFactors_WhitePaper.pdf, viewed June–August, 2017.

Isatto, E., Azambuja, M., & Formoso, C. (2013). The role of commitments in the management of procurement make to order chains. *Journal of Management in Engineering, 31*(4).

Kepner Tregoe. (2017). *Home page*. www.kepner-tregoe.com/about-kt, viewed March–September, 2017.

Kepner, C. H., & Tregoe, B. B. (2013). *The new rational manager*. Princeton, USA: Kepner Tregoe Inc., Publishing.

Kester, L., Hultink, E., & Griffin, A. (2013, June). Empirical exploration of the antecedents and outcomes of NPD portfolio success. In *International Product Development Management Conference, Paris, France*.

King, W., Chung, T., & Haney, M. (2008). Knowledge management and organizational learning. *Omega, 36*(2), 167–172.

Kogut, B., & Zander, I. (1993). A knowledge of the firm and the evolutionary theory of the multinational corporation. *Journal of International Business Studies, 24*(4), 625–645.

Koskela, L., & Howell, G. (2002). *The underlying theory of project management is obsolete*. Paper presented to the July 2002 PMI Research Conference, Seattle, Washington, USA, published by Project Management Institute, pp. 293–302.

Kossen, C. (2008). *Critical interpretive research into the life world experiences of mature-aged workers marginalised from the labour force* (Ph.D. thesis). James Cook University, Australia.

Ling, F. (2003). Managing the implementation of construction innovations. *Construction Management and Economics, 21*(1), 635–649.

Lingard, H., & Francis, V. (2004). The work-life experiences of office and site-based employees in the Australian construction industry. *Construction Management and Economics, 22*(9), 991–1002.

Lingard, H., Zhang, R., Harley, J., Blismas, N., & Wakefield, R. (2014). *Health and Safety Culture* (RMIT University Centre for Construction Work Health & Safety Report), pp. 1–124.

Love, P. (2006). Melbourne celebrates the 150th anniversary of its eight hour day. *Labour History, 91*.

Love, P., Irani, Z., & Edwards, D. (2004). A seamless supply chain management model for construction. *Supply Chain Management: An International Journal, 9*(1), 43–56.

Love, P., Davis, P., Edwards, D., & Baccarini, D. (2008). Uncertainty avoidance: Public sector clients and procurement selection. *International Journal of Public Sector Management, 21*(7), 753–776.

Lupton, S., Cox, S., & Clamp, H. (2007). *Which contract? Choosing the appropriate building contract.* UK: RBIA Publishing.

Manley, K. (2006). *Identifying the determinants of construction innovation.* Paper presented in Dubai Joint International Conference on Construction Culture, Innovation and Management, November 26–29, 2006.

Miller, M., Furneaux, C., Davis, P., Love, P., & O'Donnell, A. (2009). Built environment procurement practice: Impediments to innovation and opportunities. In: *Built Environment Industry Innovation Council Report.* https://eprints.qut.edu.au/27114/1/Furneaux_-_BEIIC_Procurement_Report.pdf, viewed April, 2017.

Mosman, N. (2013). Last planner. In *Lean Construction Institute.* http://www.leanconstruction.org/media/docs/Mossman-Last-Planner, viewed April, 2017.

Murphy, M., Perera, R., & Heaney, S. (2008). Building design innovation. *Journal of Engineering, Design and Technology, 6*(2), 99–111.

Myers, S., & Marquis, D. (1969). *Successful industrial innovations: A study of factors underlying innovation in selected firms.* Washington, DC, USA: Government Printing Office.

Naoum, S. (2003). An overview into the concept of partnering. *International Journal of Project Management, 21*(1), 71–76.

Naoum, S. G., & Egbu, C. (2016). Modern selection criteria for procurement methods in construction. A state-of-the-art literature review and a survey. *International Journal of Managing Projects in Business, 9*(2), 309–336.

National Research University—Higher School of Economics (NRUHSE). (2013). *Innovative construction materials and technologies: Their influence on the development of urban planning and urban environment* (World Experience—Research Report). Moscow: The Russian Federation, NRUHSE.

Nowotarskia, P., Pasáawskia, J., & Matyjaa, J. (2016). Improving construction processes using lean management methodologies—cost case study. *Procedia Engineering, 161,* 1037–1042.

Office of Chief Economist. (2015). The impact of regulation on Australian businesses. In *Australian Government.* https://industry.gov.au/Office-of-the-Chief-Economist/Events/Documents/8-Abrie-Swanepoel, viewed April–June, 2017.

Organisation for Economic Co-operation and Development. (2017). *Economic surveys—Australia.* https://www.oecd.org/eco/surveys/Australia-2017-OECD-economic-survey-overview.pdf, viewed July, 2017.

Owen, J., & Burstein, F. (2005). Where knowledge management resides within project management (Chap. IX). In M. Jennex (Ed.), *Case studies in knowledge management.* USA: Idea Publishing Group.

Pugh, S. (1991). *Total design: Integrated methods for successful product engineering.* Boston: Addison Wesley Publishers.

Raheem, A., & Hinze, J. (2014). Disparity between safety standards—A global analysis. *Safety Science, 70,* 276–287.

Reason, J. (2000). Safety paradoxes and safety culture. *Injury Control and Safety Promotion, 7*(1), 3–14.

Ribeiro, F., & Ferreira, V. (2010). Using knowledge to improve preparation of construction projects. *Business Process Management Journal, 16*(3), 361–376.

Ritchie, M. (2016). State of waste 2016—Current and future Australian trends. In *MRA Consulting Newsletter,* April 20. https://blog.mraconsulting.com.au/2016/04/20/state-of-waste-2016-current-and-future-australian-trends/, viewed March–April, 2018.

Ritchie, M. (2017). State of waste data (2017). *MRA consulting newsletter,* https://blog.mraconsulting.com.au/2017/03/29/the-state-of-the-waste-data/, viewed March–April, 2018.

Safe Work Australia. (2012). Australian work health and safety strategy 2012–2022. In *Safe work Australia.* https://www.safeworkaustralia.gov.au/about-us/australian-work-health-and-safety-strategy-2012–2022, viewed April–June, 2017.

Sayfullina, F. (2010). Economic and management aspects of efficiency rise of innovative activity at building enterprises. *Creative Economy, 10*(46), 87–91.

Schumpeter, J. A. (1947). The creative response in economic history. *Journal of Economic History, 7,* 149–159.

Slaughter, E. (1993). Innovation and learning during implementation: a comparison of user and manufacturer innovations. *Research Policy, 22*(1), 81–95.

Slaughter, E. (1998). Models of construction innovation. *Journal of Construction Engineering and Management, 124*(3), 226–231.

Sowards, D. (2004). 5 S's that would make any CEO happy. In *Contractor Magazine,* February 2004.

Suprun, E., & Stewart, R. (2015). Construction innovation diffusion in the Russian Federation. Barriers, drivers and coping strategies. *Construction Innovation, 15*(3), 278–312.

Tatum, C. B. (1987). Innovation on the construction project; a process view. *Project Management Journal, 18*(5), 57–68.

Tatum, C. B. (1991). Incentives for technological innovation in construction. In L. M. Chang (Ed.), *Preparing for Construction in the 21st Century 1991, Proceedings of the Construction Conference, ASCE* (pp. 447–452). New York, USA.

Teece, D., Pisano, G., & Shuen, A. (1997). Dynamic capabilities and strategic management. *Strategic Management Journal, 81*(7), 509–533.

Toner, P. (2006). Restructuring the Australian construction industry and workforce: Implications for a sustainable labour supply. *The Economic and Labour Relations Review, 17*(1), 171–202.

Townsend, K., Lingard, H., Bradley, L., & Brown, K. (2011). Working time alterations within the Australian construction industry. *Personnel Review, 40*(1), 70–86.

Tully, S. (2016). Meet the private company that changed the face of the world. In *Fortune Magazine,* May 17.

Unisearch Ltd. & Australian Royal Commission into the Building and Construction Industry. (2002). *Workplace regulation, reform and productivity in the international building and construction industry.* Paper prepared on behalf of Unisearch Ltd, University of New South Wales for the Royal Commission into the Building and Construction Industry Melbourne. http://www.royalcombci.gov.au/docs/Complete%20Discussion%20Papper%2015.pdf.

Voss, G. B., Sirdeshmukh, D., & Voss, Z. (2008). The effects of slack resources & environmental threat on product exploration and exploitation. *Academy of Management Journal, 51*(1), 147–164.

Welch, L. (2009). Improving workability in construction workers – let's get to work. *Scandinavian Journal of Work, Environment & Health, 22*(3), 321–324.

Williams, A., Franche, R., Ibrahim, S., Mustard, C., & Layton, F. (2006). Examining the relationship between work-family spillover and sleep quality. *Journal of Occupational Health Psychology, 11*(1), 27–37.

Wing, L., & Chua, A. (2005). Knowledge management and project abandonment: An exploratory examination of root causes. *Communications of AIS, 16,* 723–743.

Wood, J., Zeffane, R., Fromholz, R., Morrison, R., & Seet, P. (Eds.). (2012). *Organisational behaviour: Core concepts and applications.* USA: Wiley.

Work Health Victoria. (2013). *Work health checks in Victoria's construction industry.* Worksafe Victoria.

Yi, W., & Chan, A. (2014). Optimal work pattern for construction workers in hot weather: A case study in Hong Kong. *Journal of Computing in Civil Engineering, 29*(5).

Yin, R. (2003). Case study research—Design and methods. *Applied Social Research Methods, 15.*

Zero Waste Scotland. (2017). *The cost of construction waste,* Government of Scotland. http://www.resourceefficientscotland.com/content/cost-construction-waste, viewed February–June, 2017.

Zhuge, H. (2006). Knowledge flow network planning and simulation. *Decision Support Systems, 42*(2), 571–592.

Chapter 3
"Short-List" of Four Methodologies

3.1 Option Selection Background

The four selected streams chosen after completion of the KT DA process form the "short-list" of options, from which a "Go Forward" option(s) will be selected for further research, after an extensive "short-list" literature review. This risk-based DA framework shall also be utilized in the selection of the "Go Forward" option(s) for detailed research, which form the basis of this book (Kepner Tregoe 2013). The "short-list" of options that have the potential to enhance Australian construction engineering industry efficiency and effectiveness is: (a) Optimal Work Duration on Site; (b) Construction Site Waste Reduction; (c) Rationalization of Australian Construction Safety Regulations; and (d) Sustainable Construction Labour Force.

These four options were researched in the literature to an extent that allowed for sufficient information to prepare a briefing paper for an independent peer review panel to use when undertaking the Kepner Tregoe risk-based analysis to decide the "Go Forward" case, which will subsequently be researched in depth, during the third and final phase of this study. The following is the literature review of these four cases, building on the acquisition data base developed from the preliminary research of the literature summarized in Chap. 2.

3.2 Four-Option "Short-List"

Section 3.2 provides the initial literature review of four methodologies, interchangeably known as "options" or "cases", which form a "short-list" of potential methodologies that will potentially improve the effective and efficient functioning of the Australian engineering construction industry.

© Springer Nature Switzerland AG 2019
P. G. Rundle et al., *Effective Front-End Strategies to Reduce Waste on Construction Projects*, https://doi.org/10.1007/978-3-030-12399-4_3

3.2.1 Option A "Short-List"—Optimal Work Duration on Site

A basic efficiency criterion for executing construction work with optimal productivity is the determination of the hours of work per day (i.e. one shift) and the shifts to be worked each week, along with appropriate overtime durations (Knowles 2002; Horner and Talhourne 1995). The current research indicates that an eight-hour day on construction projects, and generally for all work, is the optimal daily shift duration. There has been little or no academic literature discovered that scientifically attempts to calculate the optimum shift time beyond observation of varying durations. English socialist, Robert Owen, had raised the demand for a ten-hour day in 1810, during the industrial revolution, and instituted it in his socialist enterprise at New Lanark, Great Britain. By 1817, he had formulated the goal of the eight-hour day (Chase 2009; Illinois Labour History Society 2008). Women and children in England were granted the ten-hour day in 1847; French workers won the 12 h day after the February Revolution of 1848 (Chase 2009; Illinois Labour History Society 2008).

The Independent Society of Operative Stonemasons of Victoria, Australia, were the first workers, internationally, to be awarded an eight-hour working day by employers; it was their campaign for an eight-hour day, ending in an agreement for this shift duration, which caused an international change in worker's rights (State Library of Victoria 2017). The International Workingmen's Association took up the demand for an eight-hour day at its convention in Switzerland in 1866, stating that "[t]he legal limitation of the working day is a preliminary condition without which all further attempts at improvements and emancipation of the working class must prove abortive", and that "Congress proposes eight hours as the legal limit of the working day" (International Labour Organization (ILO) 2017, n.p.).

Marx (1867), the founder of the theory of communism, believed that the eight-hour day was critical to workers' health, writing in Das Kapital that:

> extending the working day, therefore, capitalist production... not only produces a deterioration of human labour power by robbing it of its normal moral and physical conditions of development and activity, but also produces the premature exhaustion and death of this labour power itself (Marx 1867, p. 3).

The first country to introduce the eight-hour working day, by legal statute, was Soviet Russia in 1917. The eight-hour day was the first agenda item discussed by the International Labour Organization (ILO), which resulted in the Hours of Work (Industry) Convention 1919, which was executed by 52 countries (ILO 2017).

Although there were initial successes in achieving an eight-hour day by the Australian labour movement for stonemasons in 1856, most employees waited until between the early and mid-twentieth century for the eight-hour day to be universally achieved within the industrialized world by way of legislative regulation. Pursuant to the above, there is a heuristic belief that an eight-hour day is the optimal shift duration for humankind; this goes back to the industrial revolution and is steeped in the universal labour socialist movement. Particularly in Australia, the right is seen as sacrosanct and is still celebrated as "Labour Day", with many union offices carrying the talisman like "888" symbol above the entranceway (Monuments Australia 2017).

The eight-hour week continues to be virtually a given constant as the optimal shift duration, even though 38 h per week is now the mandatory work week. However, this is covered by rostered days off that allow for the continued maintenance of the eight-hour shift on site as the standard.

The optimal shift duration on a given project must also allow for variations in geography (e.g. remote sites in a harsh climatic environment or remote urban redevelopments in Western Sydney), work culture (e.g. highly unionized environment, such as the Kwinana Industrial Zone in Western Australia, versus a non-union commercial building site in inner city Perth), as well as short- and long-term project schedule goals. Along with the relentless drive for optimum efficient productivity, which for a contractor will always entail meeting budget estimates and schedule benchmarks, there are construction employee considerations for determining optimal site hours (Dozzi and Abourizk 1993; Langston 2014; Lingard et al. 2014).

There has been an observed tendency by the author in the 12 countries where the author has worked in engineering and construction, that there is a large burnout of key management, supervisory and professional/technical staff in Australia, more so than other countries. This appears to be due to unremitting long hours on account of overhead budgetary constraints, namely insufficient management, supervisory and technical personnel resources, limiting the capacity to delegate workload down the organizational chain of command structure. When working for US contractors, the author notes there is a depth of personnel down the organizational hierarchy, which promotes delegation to the lowest capability level to complete the task. The author reasons that economies of marketplace scale allow US contractors to hold onto a wide cohort of staff because of work opportunities and more investment capital, allowing clients to mitigate risk by paying for work to be executed with the necessary supervision.

The Horner and Talhourne (1995) study on building site manual labour productivity states that there is a consistently steady corresponding loss of productivity, as working hours increase above 40 h per week. An increase by 10 h per week, to a 50 h per week duration, resulted in a 10 per cent drop in production on a building site. For an increase to a 60 h work week, productivity plummeted by 20 per cent. When the working week was then increased up to 70 h, productivity fell by 30 per cent, on the case study site.

Knowles (2002) argues that there are an optimum number of manual workers to perform any construction site task and if numbers exceed that level, productivity will reduce; a statistical analysis was undertaken to prove this. When worker crew numbers are increased by 20 per cent above optimum task requirements, productivity reduces by eight per cent; the productivity drop was 15 per cent, with a crew size that was 40 per cent above optimum task requirements. By doubling ideal manual staff numbers on a particular site task, Knowles (2002) suggests productivity decreases by a staggering 30 per cent.

Coupled with the leadership, "burnout" factor is the tendency to increase site hours in response to lagging schedule issues for manual staff (Knowles 2002). There have been insufficient specific studies carried out on researching construction site worker accidents versus injury/serious accident/fatalities. However, there are a number of papers that allude to this correlation (Dembe et al. 2005). Without pre-empting research approaches until further detailed literature review is undertaken, parameters would need to be set that define optimum employee hours on construction sites to achieve, say staff work/life balance, optimal productivity and maximized worker safety. On the limited literature review undertaken, there is a *prima facie* case, to undertake, or as a minimum to establish, the boundaries for further research in this area.

In conclusion, an increase in manual labour and working longer hours to accelerate construction works can actually destroy productivity and negatively impact site morale of both manual and professional non-manual staff. A mathematical model could be developed for use on construction projects, which would account for suitable productivity levels, but allow for the effective theoretical attainment of project benchmarks, through striking a revised optimal shift duration for discrete periods, or the entire project life, considering crew and plant requirements. That model could set a work/life balance constant (by off-shift minimum durations) and a maximum shift time that historically delivers lowest safety incident outcomes.

3.2.2 Option B "Short-List"—Construction Site Waste Reduction

Ritchie (2016) states that Australia's population in February 2016 was 24 million people, and from 1996 to 2015, our population rose by 28 per cent. However, during this period, waste generation increased by 170 per cent and in 2016 it was estimated that 50 million tonnes of waste was generated in Australia, with construction and demolition rates being consistently around 40 per cent of total waste volume (Ritchie 2016; Waste Management World 2016). Waste generation rate is a function of population growth, urbanization and per capita income; Australia's alarming annual waste increase rate is 7.8 per cent, per annum. However, in some states, such as New South Wales, with high recycling rates, landfill storage rates have been levelling off since 2005 (Ritchie 2016). Refer to Fig. 3.1, which provides details of population versus waste, year by year.

Chandler (2016) summarizes that the construction industry is the globe's major consumer of raw materials internationally, but around only one-third of construction and demolition waste is recycled or re-used, as per a recent World Economic Forum report. Similar to Australia, about 40 per cent of all solid wastes produced per annum in the USA come from construction and demolition waste. Australia is one of the ten highest producers of solid waste in the OECD (Chandler 2016). Residential housing accounts for 38 per cent of international construction waste volume; transport, energy and water infrastructure, for 32 per cent construction waste; institutional and commercial buildings, for 18 per cent of construction waste; and industrial sites, 13 per cent of global waste volume (Chandler 2016).

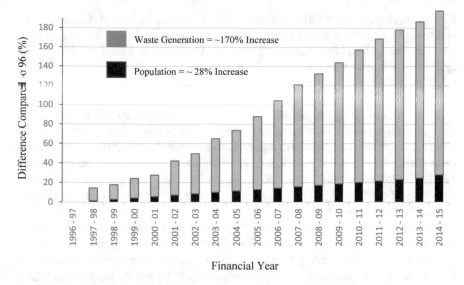

Fig. 3.1 Comparison of waste generation and population growth. Adapted from MRA Consulting Group, October 2015

Hickey (2015) argues that Australia is falling behind European recycling rates for construction and demolition (C&D) waste material, with 42 per cent of Australian C&D waste typically still going to the landfill and the balance of 58 per cent being recycled/re-used. He also suggests that government statistics quoted for construction waste treatment and landfill are often inaccurate and rely on surveys from waste levy-funded consultants, who quote each other's misinformation (Hickey 2015). Current European Union (EU) development fund advice notes that approximately 70 per cent of EU C&D waste is recycled, totalling 25–30 per cent of all wastes generated on average, which are much lower than the 40 per cent of C&D waste to total waste volume rates that typically occur in the USA and Australia (European Union (EU) 2017).

Hickey (2015) argues that because much of construction waste is handled by private landfill and recycling facilities, it is difficult to obtain accurate data; he has found that many private and indeed public landfill sites for construction waste do not have weighbridge facilities or containment linings. Federal and state governments are relying on these often inaccurate landfill figures to form government policies on construction waste to address a situation in which it is estimated that all current landfill sites, Australia-wide, will be filled by 2025.

There is a public acceptance of waste recycling as benefiting the broader community, with 51 per cent of Australian household waste being recycled, compared to 42 per cent, on average, in the EU (Planet Ark 2018). Further, this research has shown a preoccupation with engineering construction industry participants in this survey with site waste recycling, and this is borne out by a claim that construction waste recycling/re-use recovery rates in one Australian state, South Australia, is as high as

80 per cent (Zero Waste South Australia 2013). However, the overall Australian state and territory average recovery rate is 58 per cent, which is lower than the EU's 70 per cent average (EU 2017; Hickey 2015). The Australian government's 2016 "*State of the Environment Overview*", tabled to parliament in March 2017, was a follow-on report from the previous 2011 document (State of the Environment Overview (SOE) 2016).

Though the 2016 report indicated that national recycling rates had improved over the past several years, a detailed review of the data showed that the report used recycling data from states/territories that was as old as 2010/11 (SOE 2016). Outdated and inaccurately benchmarked data makes it difficult for governments and other interested stakeholders to make informed decisions on waste management. The Federal Government needs to expediently address this matter and begin producing accurate total and C&D waste data.

Furthermore, although Australian governments have political goals to improve construction waste recycling levels to world best practice standard, they have no real plans beyond the "blunt" instrument of increased levies; nor are they developing alternative strategies beyond current recycling processes to produce long-term plans to reduce construction waste volumes to landfill. All state governments (except Western Australia, who sets aside 25 per cent of its waste levy revenue for waste management improvements) are diverting often exorbitant construction waste landfill levies into consolidated revenue, away from recycling improvements, despite recycling targets consistently falling behind government criteria (Hickey 2015). Even with improving recycling rates by the construction industry over the past decade, construction and demolition waste is still the largest source of industrial solid landfill in Australia at 33 per cent (Hickey 2015). Hickey (2015, n.p.) notes that:

> [t]he levy-funded educational focus is commonly targeted at household waste via Municipal Solid Waste politically lead initiatives. These don't (sic) really assist in any change to our construction industry's landfill diversion rates (a shame, given that the construction industry is considered the largest waste generation stream on the planet).

Clearly, new approaches to construction waste need to be developed in Australia, as current processes are simply unsustainable.

Chandler (2016) estimates that approximately 3.74 million tonnes of construction waste is generated annually on Australian dwelling construction, which amounts to a cost impost of A$12,750 per dwelling for wasted material costs, site delivery, handling and off-site removal and tip fees. This amounts to a cost burden of A$2.8 billion per year, which does not include the extra carbon dioxide emissions or the wasted negative gearing subsidy footed by taxpayers.

> An evaluation on waste from a life cycle view point, would show that construction waste costs can be even higher as they not only include material costs, but also inefficiencies arising from poor planning, design, construction, operation, recycling and eventually, demolition (Chandler 2016, n.p.).

The Australian pre-build sector estimates that less than five per cent of building construction embraces prefabrication methods, and Chandler (2016) argues that these fast-track, off-site "methodologies" will greatly reduce the large volume of waste in situ construction creates at the front end. He comments on the need for a better approach to construction waste management via "the construction industry avoiding waste in the first place, though 'back-end' supply chain waste measures, such as waste recycling, will always be important" (Chandler 2016, n.p.). Jaillon et al. (2009) argue that there is a 52 per cent saving in material waste using prefabrication techniques.

Ainch and Daniel (2013) suggest that 21–20 per cent cost over-runs on construction projects were directly attributable to material wastage. Simple solutions are required, and prudent planning is necessary, to the issue of construction waste control. Professor David Chandler of Western Sydney University states that:

> This is not to say these backend waste measures are not important. Waste is still a huge challenge and minimizing its impact will always require new innovations. But the better approach for the construction industry must be avoiding waste in the first place (Chandler, 2016, n.p.).

New South Wales charges a punitively high levy of A\$133.10 per tonne of waste (metro) for landfill waste to encourage recycling, while Victoria imposes a landfill levy of A\$60.52 per tonne. South East Queensland has an average levy of only A\$30 per tonne of waste to landfill, which has resulted in significant disposal of construction and other large volume waste from New South Wales and Victoria across the border to Queensland, with an estimated total of 875,000 tonnes of waste disposed in Queensland landfill sites by these other two states in 2014 and 2015. This created an extra 15,000 truck movements along the Pacific Highway (Ritchie 2016). These cross-border truck transportation waste disposal activities are causing extra pavement wear and increasing accident risk potential.

If the root cause of the explosion in the volume of construction waste resides at the front end and not at the back end with a recycling and punitive tip levy solution, then the probable drivers of construction wastage must be examined. Nagapan et al. (2012), after an industry survey, concluded that the following factors are the key causes of construction waste generation on site, in descending order:

   I. Frequent design changes (just under 25 per cent);
  II. Incorrect material storage;
 III. Poor material handling;
  IV. Effect of weather;
   V. Ordering error;
  VI. Worker error;
 VII. Poor planning and control; and
VIII. Leftover material on site (around five per cent).

A synthesis of the literature reviews for construction waste covering the preliminary option assessment for the "long-list" and "short-list" phases has shown that there is a void in the literature for the minimization of physical construction wastage on site, including: excess concrete; earthworks soils and crushed materials; bricks; plasterboard; fixtures; electrical cabling; mechanical; equipment; timber; tiles; structural steel and the like.

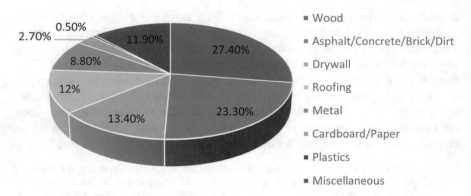

Fig. 3.2 Composition of construction and demolition waste. Adapted from Javelin (2018)

Figure 3.2 provides a construction industry average of the composition of construction and demolition waste in the USA (Javelin 2018).

The treatment of waste, matching various generally available solutions to the front-end factors contributing to construction waste, ranging from frequent design changes to leftover site material, needs to be further researched. Further literature review could be narrowed down to finding the academic and/or professional literature surrounding an OECD country with a similar construction waste profile to Australia, with a suitable research methodology that could address construction wastage mitigation in Australia.

3.2.3 Option C "Short-List"—Rationalization of Australian Construction Safety Regulations

Raheem and Hinze (2014) argue that a set of construction industry international standardized regulations are necessary because the legal framework for reporting accidents/fatalities varies dramatically between countries. Hence, worldwide comparisons of safety performance are difficult to make. The paper proposes the development of a worldwide set of construction site safety standards with limited variations to address geographical limits and economic distinctions. Construction safety statistics were collated, and methods of measurement were aligned, in order to develop a comparison of construction site accidents/fatalities in the developing world, in the BRICS economic block and in OECD countries (Raheem and Hinze 2014). The following countries were evaluated: Egypt, Tunisia, Zimbabwe, Bahrain, Hong Kong, Japan, South Korea, Costa Rica, USA, Argentina, Brazil, Bulgaria, Hungary, Italy, Portugal, Romania, Spain, Turkey, England and Australia. The safety data was collected using the following three methodologies from the International Labour Organization (tending to be old and not construction-specific) in-country labour department statistics and by a survey of in-country safety regulators (Raheem and Hinze 2014).

A comparison was then made of international safety performance between all participant countries, with an average figure calculated from 1999 to 2008 (Raheem and Hinze 2014). The data from Africa was deemed insufficiently accurate to use for comparison purposes, so not all countries in that continent were cited. Hong Kong recorded the highest figure in Asia at 60.53 fatal accidents per 100,000 working hours; Turkey topped the European figures at 37.32 per 100,000; the country in the Americas with the highest recorded safety fatalities was Argentina, with around 40 fatalities per 100,000 h worked; Australian figures, which were the lowest of all OECD countries trialled (including USA, Canada and England), were 5.7 fatal injuries per 100,000 h worked (Raheem and Hinze 2014, pp. 279–281).

Clearly, this report establishes that Australia is a world leader in construction safety. Notwithstanding this, though the Australian data showed a sharp declining trend in the fatality rate after 2001 continuing until 2005, the trend showed slight changes until 2008. A sharp rise in fatal injuries occurred during 2009. However, both a recent RMIT University report commissioned by the peak body for major Australian construction contractors (the Australian Constructors Association) and the Australian Federal Government's safety regulator (Safe Work Australia) believe that the safety performance of the Australian construction industry needs to be considerably improved (Lingard et al. 2014; Safe Work Australia 2012). In fact, 211 Australian construction employees died of work-related injuries between 2007 and 2012 and there were 13,735 construction worker serious compensation claims in that period (Safe Work Australia 2012, p. 4).

Statistically, the construction site is one of Australia's most dangerous workplaces. The RMIT University (2014) report establishes that the change process to introduce a culture of safety in the Australian construction industry, as opposed to simply a rigid adherence to a stringent regulatory framework, will lead to continuous improvement in safety and save lives. The Safe Work Australia (2012) paper also recognizes the value of changing the broader community attitude to embrace a safety culture and states that:

> [o]rganizational cultures can be influenced by broader community values and attitudes. Community expectations can be powerful drivers of change and collectively influence the nation's health and safety culture. When the Australian community expects and demands that work be free from harm, any failure to do so generates community pressure and action (Safe Work Australia, 2012, p. 9).

Bahn and Barratt-Pugh (2014) outline the need for government regulators to better understand the requirements of industry trainers (who, in their opinion, form the backbone of transferring the relevant safety act requirements to construction workers) and the need for governments to fully integrate construction safety standards throughout the country; currently, some states have maintained their own regulatory framework and yet several states adhere to a national safety construction standard. However, there is another line of academic thought that the best safety outcomes can be achieved by project leadership working on reinforcing site personnel positive safety behaviour (Chowdry 2014).

Australia was ranked 128 out of 148 countries in how burdensome the government was for obtaining permits, regulations and statutory reporting requirements, as part of the 2014 Global Competitive Report. This may in part explain why Australia dropped out of the top 20 countries, competitive business-wise, for the first time in 20 years and was rated 21st in the world; this was the first occasion that Australia received a lower rating than New Zealand (ABC 2014; World Economic Forum 2014).

The *"Australian Government Taskforce on Reducing Regulation Report"* (Commonwealth 2006) recognized that although Australia would not function effectively without regulation, Australian business is being negatively impacted by over-regulation which is driving down efficiency. The task force report argues that as a nation, Australia has become increasingly wealthy and correspondingly risk adverse, to the extent that its citizens expect governments to protect them from inherent risks of daily life ranging from loss of life to loss of possession. These regulatory imposts are causing considerable loss of profit, as exemplified by the Australian Consulting Engineers of Australia, which reckoned that unnecessary regulations were burdening the profession by more than A\$18 million per annum. This 2006 Task Force Report made a number of constructive recommendations surrounding Australian occupational health and safety regulations to assist streamlining these processes without increasing the worker risk profile (Commonwealth 2006).

Although there is a raft of data from academic, government and trade organizations on safety regulations in the Australian construction industry, there is a need to further evaluate any potential to rationalize these statutory laws. In conclusion, it stands to reason that a thorough review of construction industry safety standards and regulations is necessary, with a view to optimizing efficiencies while maintaining effective safety outcomes, to evaluate appropriate methodologies to achieve these goals. This research could result in commercial benefits to contractors, along with the community at large via public project savings, and also be of social benefit via a reduction in accidents and injuries. Any assessment as to a path forward to effect a cultural change in safety in the construction industry would be outside the limited scope of this.

3.2.4   Option D "Short-List"—Sustainable Construction Labour Force

Section 3.2.4 evaluates the issues surrounding the development of a long-term sustainable construction workforce in Australia to robustly cope with the country's often disruptive economic cycles that are more and more impacted by external drivers of change in international market forces, such as the GFC, Brexit and the US dollar exchange rate, along with the impact on Australia's export market via international commodity prices (OECD 2017). Around 1.05 million personnel are engaged in the Australian construction industry, which variously represented around nine per cent of all employees in mid-2015; only the healthcare and retail industries employ more people (AGDE 2015). Despite job losses in heavy engineering construction due to the resource commodity price slump, in 2014 and 2015, the total construction labour force in Australia has increased due to residential housing and apartment construction, particularly in New South Wales and Victoria.

Overall construction numbers were down in Queensland, South Australia and Tasmania in the five years leading up to 2015, but these construction employment numbers increased during that five-year period in the largest two states, New South Wales (11 per cent at 30,000 personnel) and Victoria, along with Western Australia, Northern Territory and Australian Capital Territory (ABS 2015a). There has been a large swing into residential and commercial building, and the fabric of the construction workforce has altered from heavy construction to a construction labour pool dominated by 23 per cent of personnel engaged in "building installation services" (232,700 workers) (ABS 2015a). Notable employment and employee characteristics in this industry include far higher shares of full-time, younger, self-employed, certificate-qualified male workers than for most other industries; 85 per cent of construction workers are full time. Only 11.4 per cent of the male-dominated construction labour force are women, compared to the industry-wide average of 45.7 per cent (AGDE 2015).

Older workers are under-represented in the construction industry; this is likely due to the manual labour roles dominant in the industry, with 35 per cent aged 45 years or over, compared with 39 per cent across all industries (The Australian Industry Group (AIG) 2015). The median age of workers in the industry was 38 years in 2014, slightly below the median age of 40 years recorded across all industries (AIG 2015). By contrast, the industry is the largest employer of young full-time workers, employing 152,100, or 17 per cent, of all full-time workers aged 15–24 years in February 2015. The most common occupations in the construction industry in 2014 were carpenters and joiners (105,900), electricians (87,900) and plumbers (73,600); 30 per cent of construction workers are self-employed, compared to only 8.5 per cent for all industries (AGDE 2015). Forty-five per cent of construction industry workers have completed a Certificate III or IV qualification. This is well above the share for all industries (20 per cent), and in contrast, only eight per cent of construction industry workers have attained a bachelor degree or higher, compared with 28 per cent for all industries (ABS 2013).

The literature review has shown that there are many areas related to creating a more Sustainable Construction Labour Force in Australia, including: further research on the poor health and life style evident amongst this construction cohort; retention of older construction workers within the labour pool to draw on their valuable knowledge and experience; and importantly, worker training and education. Another area that requires urgent further scrutiny is the Federal Government's April 2017 announcement that it is phasing out the Temporary Skilled Visa (sub-class 457 visa) which shall be replaced in 2018 by the Temporary Skilled Shortage Visa (TSS), with the purpose of providing Australian citizens with potentially improved work opportunities (Keay 2017). The TSS provides few allocations and more restrictive entry criteria for foreign applicants. There are presently only 95,000 foreign workers employed on short-term work in Australia on a 457 visa, which represents less than one per cent of our total labour force (Keay 2017).

The Australian engineering consulting and construction industries rely on rare and valuable human resources as defined in the VRIO human resources model (Barney and Hesterly 2010), which argues that employees with highly unique skill sets give companies considerable competitive advantage. In highly specialized theatres and with a total employment pool of only 12 million workers in Australia, foreign expertise (in the sciences, professionals, academia and management sectors, particularly in the engineering and construction fields, along with certain on-site trade and procurement occupations) must be drawn upon.

There is a compelling argument that the replacement of the 457 visa and with the much more rigid TSS visa will actually hamper the Australian construction industry (with many of our major construction and engineering consulting houses now being owned by overseas companies) due to restrictions being placed on the vital and healthy transfer of international intelligence. In fact, Australian unions are concerned that there will be little or no improvement in worker jobs or living standards as a result of these changes to work visas (Keay 2017).

The literature review has revealed that the most obvious means of building a more sustainable construction workforce in Australia is through the recruitment of more women into the industry. With only 11.4 per cent of the Australian construction labour force being female, there is an obvious gender resource that needs to be tapped. Lingard and Lin (2003) carried out a survey of women in the Australian construction industry to evaluate the connection between career, family and work environment factors on women's organizational commitment. The report concluded that construction firms endeavouring to improve organizational commitment among female employees should ensure that women have access to career development opportunities and that equitable processes are used in providing organizational rewards.

It was also suggested that construction companies develop diversity initiatives and educate employees, particularly managers, in supportive management practices for a diverse workforce. Fielden et al. (2000) note that women constitute only 13 per cent of the UK's construction workforce, although females make up approximately half of the total UK workforce, making the construction industry, by far, the most male-dominated sector of all British industrial groups. The research found that barriers to women entering the construction industry included "the construction industry's image; career knowledge amongst children and adults; selection criteria and male dominated courses; recruitment practices and procedures; sexist attitudes; male dominated culture; and the work environment" (Fielden et al. 2000, Abstract).

A December 2016 report by the University of New South Wales (UNSW 2016) on women in the Australian construction industry posits that the construction business is the "last frontier" to be overcome for gender equality in this country. The report, which focuses on professionals who design and manage construction projects, states that around 12 per cent of the total construction workforce of one million are female; professional positions are 14 per cent women, but in the trades sector, women represent less than two per cent of site workers (UNSW 2016). Despite the construction industry implementing various mentoring and training programmes, along with paternity leave schemes, female representation is actually going backwards (UNSW 2016). Ten years ago, the construction workforce was 17 per cent female (UNSW 2016).

Women who enter the construction industry leave 39 per cent faster than men (UNSW 2016). Interestingly, the report concludes that though there are gender-specific changes that need to be made, the change management bigger actually required is a cultural shift away from the current "1950s workplace culture" of 60–70 h working weeks and consecutive 14 h shifts, which allows little time for family life and work/life balance. The report concludes that construction men, with twice the national average for suicide rates and by far the largest substance abuse rate than any other sector, have also become a victim of this outdated culture, which women are much less willing to tolerate (UNSW 2016).

In conclusion, the above-noted UNSW report (2016) recommends that the industry embraces the 40 h per week, and factors in job sharing, part-time roles and parental and care leave, to support both women and men into project schedules. It further recommends the introduction of zero tolerance for sexist behaviour and language and to make management at head office and on site responsible, as a means of increasing female participation in the construction workforce, as well as creating a more sustainable employment pool of both men and women.

3.3   "Short-List" Decision Analysis to Select "Go Forward" Case

3.3.1   "Short-List" Decision Analysis Framework

A formal Kepner and Tregoe (2013) risk-based decision analysis process was used to independently determine a "Go Forward" case for further detailed research, from the four-option "short-list" considered in this chapter.

The first step in this process was the selection of the independent peer review panel who would provide an objective decision analysis of the four options, relatively scoring each option to allow selection of the "Go Forward" case(s). This peer review panel had a different composition of construction engineering personnel to that of the "long-list" panel, which drew on a wider sector range of engineering construction industry specialists (including an environmental scientist, construction lawyer, project financier and commercial manager) to provide the best overview of the eight "long-list" options. The "short-list" peer review panel of nine academics and industry practitioners had a strong focus on design and delivery expertise. This was because the "Go Forward" case for detailed research required academic research capabilities to determine such matters as a literature void warranting research, coupled with practitioner experience in: engineering design; architectural building design; heavy construction and infrastructure construction; high-rise and commercial building; waste management service provider experience; and large project client representative expertise.

The independent facilitator, a senior SCU lecturer who also ran the previous "long-list" process, was engaged to facilitate this formal Kepner Tregoe decision analysis (KT DA) workshop and to review the risk assessments previously prepared for these options. The second step in the "short-list" decision analysis process was the preparation of an option risk assessment, which both the facilitator and the peer review panellists reviewed and indicated that the risk review was "fit for purpose"; the facilitator, Dr. J. Novak, has a Ph.D. in a risk management-related field.

The next step entailed the formulation of a DA matrix for use as a scorecard to be circulated amongst nine eminent engineering construction specialists who would form the independent peer review panel and would be given the task of reviewing the four options and scoring them on the DA matrix scorecard. A document package was transmitted via the Internet to each panellist, which included a scorecard for each option, the risk review, details on the author's assessment of the probable size of the void in the literature for each option and a briefing paper containing the literature review of each option, as found in this chapter. The panellists were asked to comment on the appropriateness of this DA matrix scorecard pro forma. Any panellist comments were addressed by the author.

Eight of the nine invited panellists attended the teleconference, with one member sending their regrets on the day. The author scored the four options prior to the workshop, but as he acted as a technical advisor during the workshop, he did not vote in these proceedings, which were scored by the facilitator. The peer review panel requested that, as the two-hour workshop was completed in such a collegiate manner, an average score for each option was calculated. This was unanimously concurred. Several questions arose from the panellists, which entailed a further period of the literature review by the author, to address these queries.

Of the four options, Rationalization of Australian Construction Safety Regulations, which was the low score, was rejected by the panel as being highly unlikely to be implemented due to Australia's high regulatory and industrial relation landscape.

During the event, the ultimate selection of the "short-list" proved the prudent choice, as a wealth of areas of possible research for this option was an outcome of further literature review on construction waste and the void in the literature was readily verified. The single "Go Forward" option was developed as: *"Front-End Strategies to Reduce Waste on Australian Construction Projects"*. It should be noted that the author had the power of veto to select a "Go Forward" case for further research different from that of the peer review panel's recommendation. However, the author's scorecard and the panel's scorecard both rated the Construction Site Waste Reduction case as the most suitable option.

The details of this Kepner Tregoe process to determine the "Go Forward" case by independent peer review panel are detailed in the following.

3.3.2 List Peer Review Panel

Table 3.1 summarizes the details of the nine members who agreed to form the independent peer review panel to determine the "Go Forward" case to be further investigated during the phase three detailed research stage. Only one member, Mr. T. Sleiman, was unable to attend the international workshop on 17 July 2017 by Skype teleconference, and advised on the day, due to an unforeseen project site

Table 3.1 Panel participant list for decision analysis

Independent peer review panel participant list for decision analysis to select "short-list"		
Panel member	Highest academic qualifications	Profession
Mr. A. Bahadori	Ph.D. (Eng.)	SCU senior lecturer with a background as a chemical engineer in oil and gas
Mr. K. Doust	Ph.D. (Eng.)	SCU engineering coordinator with a background as a civil engineer in railway transport
Mr. A. Morgan	MBA	Civil engineer with a background as a consultant branch manager for a large process and mining design consultant and mega-project management specialist for BHP Billiton
Mr. K. Bradley	B. Eng. Sc. (Hons)	QA manager for Leighton contractors for 20+ years
Mr. G. Stacey	B. Eng. Sc.	Construction manager for heavy infrastructure contractor, clough and ex-GHD design manager
Mr. D. Mulvihill	B. Arch.	Company director of architectural practice on commercial building projects with residential building experience
Mr. J. Hayton	Certificate III Carp. & Building Lic.	Site manager on four mixed commercial high-rise developments for major Melbourne developer, pace
Mr. G. Murphy	Dip. CE	Chief Executive of the Lismore City Council's Engineering Department
Mr. T. Sleiman	B. Eng. Sc.	Infrastructure project director for John Holland

By P. Rundle, July 2017

Table 3.2 Risk assessment matrix

Likelihood	Impact				
	Insignificant	Minor	Moderate	Major	Severe
Almost certain	Moderate	High	High	Extreme	Extreme
Likely	Moderate	Moderate	High	High	Extreme
Possible	Low	Moderate	Moderate	High	Extreme
Unlikely	Low	Moderate	Moderate	Moderate	High
Rare	Low	Low	Moderate	Moderate	High

Adapted from BHPB Risk (2007)

issue. Because the KT DA indicated a clear first choice option, and since the author's assessment concurred with this option, Mr. Sleiman's scorecard would not have impacted the "Go Forward" selection. The facilitator advised the panel of this fact after calculating the result scorecard summary, which can be found in Appendix B at the back of this document.

3.3.3 "Short-List" Option Risk Assessment

Table 3.2 details the risk assessment undertaken for each of the four "short-list" items, based on three identified key potential risks related to construction engineering innovation processes. Table 3.2 risk assessment was validated by the Kepner Tregoe facilitator and peer review committee panellists as suitable for use. See Sect. 2.3, ""Long-List" Decision Analysis to Select "Short-List"", for the full details of the development of these Table 3.2 option risk assessments from the first principles. Note that the Kepner Tregoe facilitator has a Ph.D. in a risk management field (Table 3.3).

3.3.4 "Short-List" Decision Analysis Score Card

Table 3.4 is a pro forma only of the score used by the peer review panel.

The formal Kepner Tregoe scorecard has been adapted from the previous "long-list" scorecard, but now includes "must have" and "want" criteria. Failure of an

Table 3.3 "Short-list" risk assessment of "long-list" options

Option no.	Option	Market demand risk rating	High capital cost risk rating	Budget blowout risk rating	Overall risk rating	Remarks
A	Optimal Work Duration on Site	x	x	x	Low	Low-risk approach with high payback potential
B	Construction Site Waste Reduction	x	x	x	Low	Low-risk approach with high payback potential
C	Rationalization of Australian Construction Safety Regulations	x	x	x	Moderate	There could be resistance from government regulators to rationalize the status quo that would dilute demand appetite
D	Sustainable construction labour	x	x	x	Moderate	Considerable public funding required to promote industry training and there is entrenched bias about perceived age and gender matters

By P. Rundle, 2017 (Adapted from BHPB Risk 2007; Kepner and Tregoe 2013; Pugh 1991)

Table 3.4 Scorecard for decision analysis of four-option "short-list" to prepare "Go Forward" case

Option no. option description DA scorecard for determination of "Go Forward" case(s)

No.	Option marking criterion	A Marking criterion weighting	B Peer panel member option score	A × B Option criterion total
Kepner Tregoe "want" marking criteria for each option				
1.	Positive impact of option on efficiency defined as: *meeting all internal requirements for cost, margins, asset utilization and other related efficiency measures* in Australian construction industry	5		
2.	Positive impact of option on effectiveness defined as: *satisfying or exceeding customer needs with a compliant product* in Australian construction industry	5		
3.	Commercial benefit of option to construction industry stakeholders including contractors, suppliers, consultants and customers	4		
4.	Social and environmental benefit of option	3		

(continued)

Table 3.4 (continued)

Option no. option description DA scorecard for determination of "Go Forward" case(s)

No.	Option marking criterion	A Marking criterion weighting	B Peer panel member option score	A × B Option criterion total
5.	Ease of option application and roll-out in the construction industry	3		

Kepner Tregoe "must have" marking criteria for each option

No.	Option marking criterion	A	B	A × B
6.	Identified void in the literature per option (rating by P. Rundle)	TBA P. Rundle		
7.	Risk profile of option (rating by P. Rundle)	TBA P. Rundle		
Grand total option no. description				

By P. Rundle, 2017 (Adapted from Kepner and Tregoe 2013)

option to pass the "must have" criterion precluded it from further consideration. This change to the pro forma was made to align with the formal process utilized for this assessment, which also included the requirement for the proceedings to be convened by a facilitator and for the panellists to attend "live" in a workshop environment.

3.3.5 "Short-List" Decision Analysis to Determine the "Go Forward" Case(S)

Dr. J. Novak, a specialist in risk management and critical infrastructure, independently facilitated the 17 July 2017 "Go Forward" case workshop, attended via Skype by the eight available peer review panel members, to select the "Go Forward" case. The author attended this workshop to address any technical questions but did not participate in option selection. The Deputy Mayor of Lismore City Council and SCU lecturer, Mr. G. Battista (Masters of Business Administration (MBA)), also attended the workshop as a "process observer". Ms. C. Roberts took key meeting point notes on behalf of the facilitator during this workshop.

The author commenced the workshop with introductions, and before handing over the floor to the facilitator, it was confirmed that only one option would go forward into the final detailed research stage, phase three. The facilitator also introduced the administrator, Ms. Roberts, who would take salient notes during the process.

The first step in the process was a validation by all attendees at the workshop that the risk assessments and pro forma decision analysis scorecard were "fit for purpose", and these details, coupled with the literature review for all four options, were provided to attendees the following week. The method of scoring was readily accepted by the facilitator and the author, as per the request of two panellists that the score for each criterion was consolidated as an average. The author then ran through each of the four options for the workshop attendees.

The facilitator then explained the Kepner Tregoe decision analysis process (KT DA) and advised that the peer review panel needed to provide a score of one to five into the option score in Column B of the KT DA scorecard, for items one to seven inclusive, for all four options, five being the highest score and one being the lowest score. Table 3.3 is a "typical" KT DA pro forma scorecard, which was provided for each of the four options.

The author, having undertaken a progressive literature review of the four considered options, was best placed to provide a score for the scorecard criterion "identified void in the literature" for each option, which is noted as follows:

I. Optimal Work Duration on Site (five—at this stage of the literature review, an apparent significant void);

II. Construction Site Waste Reduction (five—at this stage of the literature review, an apparent significant void);

III. Rationalization of Australian Construction Safety Regulations (three—large quantity of the literature but limited on rationalization issue); and

IV. Sustainable Construction Labour Force (three—significant information, however, it is disjointed).

The facilitator then went through the risk assessment for each of the options, and all of the participants agreed that the minor and moderated risk ratings, detailed in Table 3.2, did not preclude any option from further review. The facilitator then checked the two "must have" criteria and found that there was no cause to preclude these options from further assessment and that the nuance between options, presented by the "void in the literature", was picked up in the author's scoring of this criterion.

The panel members requested that they assess the scoring, one criterion at a time, in order to best obtain a comparative reference between each option. The panel also asked that the eight member's scores were added and an average derived. All peer review panel attendees proactively participated in the scorecard assessment for each of the four options. Details were recorded by Dr. Novak, with scores calculated as an agreed average by the group, rather than individually. Refer to Appendix B, "Kepner Tregoe Decision Analysis 'Short-List' Scorecard—Options A, B, C and D", which summarizes the sum of scores of each of the eight panellists, for each of the eight "long-list" options, including the scores of the author. Refer to Table 3.5, which summarizes the four-option scores.

The author's scorecard ranking concurred with the independent Kepner Tregoe assessment of the option ranking order. There was a firm indication by the peer review panel that the option that received the lowest ranking, Rationalization of Australian Construction Safety Regulations, should be discarded as a "non-workable" option,

Table 3.5 "Go Forward" option DA ranking

"Short-list" DA scores and option rankings			
No.	Option	Grand total Each option	Option Rank
IV	Construction Site Waste Reduction Low risk	117	1
V	Optimal Work Duration on Site Low risk	112	2
VI	Sustainable Construction Labour Force Moderate risk	90	3
VII	Rationalization of Australian Construction Safety Regulations Moderate risk	82	4

By P. Rundle, July 2017

even though the Commonwealth Government taskforce on regulation tabled a report in 2006, entitled *"Rethinking Regulation"* (highlighted in the literature review), that recommended OH&S regulations should be harmonized and streamlined across all nine Australian governments, as a result of a strong representation from Australian business, including the construction industry.

The peer review panel noted that Australia was still highly unionized in the construction industry, that another point made on safety in the literature review was Australia's excellent safety record compared to other OECD members, and that *prima facie*, it would be difficult to mount a challenge to the status quo by further research of this kind.

The facilitator suggested that it was opportune to revisit the KPIs for option selection, as a "sanity check". These six KPIs were noted to the peer review panel as follows:

I. The textbook will be of transparent commercial benefit to contractor/engineering houses, client end users and the community at large.

II. The proffered solutions can be readily and, therefore, expediently implemented.

III. The topics selected for investigation shall provide maximum stakeholder benefits.

IV. The solutions to the identified inefficiencies and ineffective practices are made available within the academic and professional international body of knowledge.

V.   The textbook must be practical in nature, address a void in the Australian engi-
     neering construction business, and the work must be valuable to this industry.
VI. Finally, the identified research broadly complies with a triple bottom line phi-
     losophy and provides commercial, social and ecological benefits.

The peer review panel reviewed each option against the six KPIs and was unan-
imous in their assessment, after considerable discussion. The panel noted that Construction Site Waste Reduction, Optimal Work Duration on Site and Sustainable
Construction Labour Force were all compliant with the KPIs. However, there was
robust discussion on the option which would further evaluate a Rationalization of
Australian Construction Site Safety Regulations. The final opinion of the peer review
panel was that this site safety option would not pass two KPI hurdles as: (a) it would
not be readily implemented, and (b) this option is not practical in nature given the
political, union and regulatory safety landscape. Accordingly, this option would not
be recommended for future research at a later date.

Refer to Appendix B at the end of this book, which provides the detailed Kepner
Tregoe decision analysis scorecard from which summary Table 3.5 was derived.

The peer review panel noted that although the Construction Site Waste Reduction
option was the highest scoring result, the case for the Optimal Work Duration on
Site option, which scored second, needed further clarification before it could be
vetoed. The peer review panel had several questions surrounding these options as
a result of this workshop that they requested to be addressed before advising their
recommendation and the author readily agreed.

The floor was opened at the conclusion of this workshop, and it was stated that
there were three remaining questions asked during this workshop by panel members,
which needed to be addressed before their scorecard recommendation on the Construction Site Waste Reduction "Go Forward" case could be verified. The workshop
secretary, Ms. Roberts, noted as follows:

I.   Mr. Kevin Bradley indicated that only the female gender area was researched at
     length in the brief provided on developing a Sustainable Construction Labour
     Force, and that he would be interested in an evaluation on other facets of this
     issue being considered further, such as overseas workers and retention of older
     workers. This may alter the ranking.
II. Mr. Jarryd Hayton noted that if the first two ranked options were going to be
     researched in detail, then it would be good if the author could also undertake
     further literature review on the Optimal Work Duration on Site option to define
     what the next level of research would entail and whether these outcomes would
     satisfy the Australian commercial building industry as having "value".
III. Dr. Bahadori suggested that the author needed to assure himself that he had
     a passion for the Construction Site Waste Reduction option and could look
     at verifying this by further investigating all possible topic scopes for detailed
     research through further review.

The author thanked the peer review panel for their critique as well as their workshop attendance. He undertook to carry out further two weeks of the literature review

to address Questions 1–3 inclusively and would only proceed after his co-authors were satisfied that these workshop queries were addressed. Refer to Sect. 3.3.6 for the responses.

3.3.6 Address Outstanding Peer Review "Go Forward" Workshop Questions

Responses to the three questions asked by the independent peer review panel, on 17 July 2017, were formulated by the author before the "Go Forward" option could be validated. Further literature reviews addressed these panel queries, and the author's responses are noted as follows:

3.3.6.1 Workshop Panel Question No. 1

The author to undertake further literature review on Option D, Sustainable Construction Labour Force, to consider other facets such as older workers and foreign workers and to determine if this information could influence the "Go Forward" selection.

A 2015 report by the UK Chartered Institute of Builders (CIOB 2015) was conducted on maximizing the utilization of older construction workers in Great Britain because that countries' population, in common with other OECD countries, sees a rapidly ageing population demographic. With 19 per cent of the British construction workforce looking to retire in the next five to 10 years, and a predicted industry crisis that a mooted shortage of 224,000 construction workers by 2019 would cause, it is crucial to look towards retaining older construction personnel (CIOB 2015). "Crucially, the report suggests that more needs to be done to make better use of ageing workers' expertise and skills, and use this to help up-skill younger counterparts" (CIOB 2015, n.p.). Bridget Bartlett, Deputy Chief Executive of the CIOB, noted that:

> CIOB research showed that skills shortages in construction are compounded by those entering the industry not being suitably qualified for the position and the construction industry needs to use this opportunity to use older workers to tap into their skills and knowledge and ensure they are passed onto the next generation (CIOB 2015, n.p.).

A landmark, frequently cited, 2013 study of older construction workers was carried out by the Building and Construction Industry Training Organization (BCITO), which is a New Zealand government organization and is the largest provider of construction trade apprenticeships in New Zealand. The BCITO was appointed to develop and implement industry qualifications for the building and construction sector (BCITO 2013). The following conclusions were derived from this report:

I. Employment tenure is a critical aspect for the retention of older construction workers. However, the older tradesperson cohort is the highest recipient of New Zealand's welfare/healthcare system for workplace injuries. Despite the construction industries professed desire to retain older workers because of their enormous knowledge base, the financial cost of making the workplace environment less hostile for the older workers, coupled with the ease of employment of younger migrant workers, is considered too great a burden in the highly competitive New Zealand contracting/building sector.

II. OH&S regulations need to be evaluated and amended, to enable older workers to remain in the industry for a longer duration, via targeted interventions to ensure all workers are following the act(s) of using correct tools, equipment and processes; this will extend the working life of employees. A culture of employee acceptance of injury and ill health in construction needs to be altered, and trainers, along with safety officers, should ease the physical burden of work through task evaluation analysis.

III. The BCITO believes that a coordinated response is needed by both employers and government to mitigate the pressures that work against attracting and keeping older workers (BCITO 2013).

A report executed by the University of South Australia on behalf of the South Australian State Government (Lundberg and Marshallsay 2007) on retention of older workers in the construction industry noted that in 2003, the Australian Bureau of Statistics found that 29 per cent of the Australian construction employee workforce were 45 years of age and over. This University of South Australia report concluded that there was a significant degree of age discrimination in the local construction industry towards its older workers, which needs to be addressed, and that superannuation and work cover policies discriminate against older workers and act as disincentives for older workers to remain employed in the construction industry (Lundberg and Marshallsay 2007). Older workers see themselves as disadvantaged and not deemed "trainable" to engage in new skill sets by their construction employers and are thus not given the opportunity to embrace new technological developments (Lundberg and Marshallsay 2007).

Heaton (2017) argues that measures must be taken to account for Safe Work Australia statistics of accidents and fatalities on Australian construction sites between 2003 and 2013 that indicate that a 65-year old site worker is six times more likely to be killed on a construction project than a 34-year old worker. Although considerable safety training goes into young workers new to construction, virtually no expenditure is targeted for this ageing demographic (Heaton 2017). Similarly, workers over 55 represent only 11 per cent of the construction workforce; however, this cohort accounts for 29 per cent of fatalities on Australian sites (Heaton 2017). Moreover, 40 per cent of construction-related compensation claims are submitted by workers 55 and over and the average compensation recovery period is four weeks for a construction worker under 25 and nine weeks for the 55 and over cohort (Heaton 2017). Finally, Heaton (2017, n.p.) quotes evidence that:

the value of older workers in driving safer practices on site should not be underestimated and thanks to the experiences of many now older workers in the past, safety practices for younger workers are safer today.

The issue of older worker safety must be addressed in Australia to retain this cohort

Regarding the issue of cultural diversity in the Australian construction industry, construction has always been one of Australia's most culturally diverse industries.

Australian Bureau of Statistics figures show that 52 per cent of the construction workforce were born overseas compared to 25 per cent in the wider population, and that 39 per cent were born in a non-English speaking background country (Loosemore 2015, n.p.).

Loosemore (2015, n.p.) argues that decades of evidence show that more culturally diverse companies are more innovative and that more recent research also shows that "workforce diversity is also related to reputational advantages and higher workforce engagement and satisfaction, which in turn helps in attracting and retaining more talented employees". There is a raft of evidence in the academic literature that diverse workforces open up new market opportunities and that businesses with employees who share a client's country of birth are more likely to meet client requirements (Loosemore 2015).

Another area that needs evaluation is the education and training sector for maintaining a robust and sustainable construction workforce in Australia. From 2010 to 2015, around 12,325 apprenticeships were completed annually on average, which is down on the 30,000 per year back in the early 1990s, but a marked improvement on a low of only 8000 apprenticeship completions in 2003 (Heaton 2015). Moreover, only around 50 per cent of apprenticeships are actually completed currently; however, this figure also accounts for many construction industry apprentices who take work in their trade, rather than completing their final year. Nevertheless, construction industry projections are that, from 2015 to 2020, 76,000 extra tradespeople will be required just to fill retirements, let alone industry expansion, but that from 2010 to 2015, only 61,000 tradespeople completed their course; therefore, a shortage in tradespeople seems likely (Heaton 2015).

This potential gap in 2020 of 15,000 personnel plus extra labour to cover industry expansion could possibly be filled by foreign skilled workers. However, Australia needs to encourage its 6.5 per cent unemployed personnel, as high as 20 per cent in some areas amongst the younger cohort, to seek apprenticeships (Loosemore 2015), or use a concerted effort to reduce the number of failures to complete the indentured trade period (Heaton 2015). Detsimas et al. (2016, p. 16) argue that their research:

confirms that there is a gap in workplace training within the construction industry particularly in relation to the role-specific technical training. Construction employees tend to use informal training methods more frequently than formal methods to gain practical knowledge. However, employees are less likely to benefit from a holistic skill development that is associated with participation in balanced formal/informal training.

Accordingly, the type of training and the nature of training facilities (e.g. informal training, TAFE, online training courses) also need to be investigated for the construction business.

Accordingly, if this option was adopted as the "Go Forward" case for detailed research, this book could be aimed at synthesizing all the available literatures from government, industry practitioner, trade union and academic sources to map a path forward for the industry. An important feature of this master plan would be to develop industry capability to cope with economic and demographic shifts between sectors (viz. resources to infrastructure development and changes in residential dwellings from detached to commercial apartments and the like).

Conclusion: Option D, Sustainable Construction Labour Force

The additional literature review has clearly shown that there is considerable extra trade literature, particularly the professional and academic literature, which addresses this important topic. However, this literature is extremely fragmented, and no discernible cohesive master plan appears to have been prepared for the purposes of creating a sustainable Australian construction force. Given the important nature of this leading indicator industry to our economic health (employing approximately one million Australians), this task best resides with the Australian Federal Government to undertake such a plan. Accordingly, Option D would not be a suitable topic for this textbook.

3.3.6.2   Workshop Panel Question No. 2

The author to undertake further literature review on Option A, Optimal Work Duration on Site, to define what the next level of research would entail and whether these outcomes would satisfy the Australian engineering construction industry as having "value".

Srour et al. (2006) developed an optimization-based framework for matching supply and demand of construction labour with maximum efficiency via training, recruitment and labour allocation in order to maximize available labour for the purposes of reducing project costs and improving schedule performance. Skilled construction labour problems were identified in USA, UK, South Africa and Canada. In Australia, there is also a skilled construction labour shortage issue (Heaton 2015). Castanada (2002) addressed this issue by recommending construction workers be trained in more than one skill set—possibly a soft skill such as "management". This would maximize on-site labour utilization. Optimization techniques have been used extensively in various industries, including communications, manufacturing, transportation, health care, finance and construction.

Shift scheduling is a classic use of optimization in hotels, aviation, banks, hospitals and law enforcement offices, where, typically, a certain demand pattern alters over the course of an operating week (Srour et al. 2006). Aykin (1996) looked at a solution that would provide employee assignment to various shifts, specified by duration, employee numbers, start/finish time and break patterns. Srour et al. (2006) adopted a linear programming model as the most appropriate optimization model for their mathematical modelling technique to optimize labour on site and tested this model via several case studies.

Another relevant paper that a review of the literature has uncovered was written by Yi and Chan (2014), who developed an optimization model for the purpose of maximizing construction worker productivity in Hong Kong during the hot and humid season where extreme weather conditions are encountered. Optimal productivity was obtained via evaluating start and finish times, break times and maximizing peak working periods outside the probable extreme heat of the day window. Yi and Chan (2014, p. 1) qualify that:

> having established a Monte Carlo simulation-based algorithm to optimize work–rest schedule in a hot and humid environment, this paper attempts to develop the algorithm and identify an optimal work pattern, which may maximize the direct-work rates and minimize the health hazard due to heat stress to the workers concerned.

The developed optimization algorithm model ran through 22 work patterns, while maintaining the traditional nine hour construction site day, used in Hong Kong before the optimal start/finish/break times were derived with a maximum work rate of 87.7 per cent and minimum likelihood of worker heat distress.

Accordingly, if this option was adopted as the "Go Forward" case for detailed research, this textbook could be aimed at:

I. Endeavouring to find a suitable mathematical optimization model via further literature review that could be adapted to suit the Optimal Work Duration on Site framework, in order to develop a model for use in Australia and possibly internationally;

II. Determining the parameters for which the optimization could be used. As an example, in harsh remote environments and/or developing country scenarios, where fly in/fly out work patterns and long hours of work provisions in a developing nation would render a number of different variables to say, a Sydney-based high-rise commercial building project or a rural New South Wales Freeway Bypass project, with dissimilar labour relation agreements, varying productivity rates, different regulatory constraints and diverse climatic conditions;

III. To be of a sufficient level of sophistication to be of practical use to the Australian engineering construction industry, the actual development of the optimization model to comply with item (II) above would require a longer duration than the 12 months allocated to this research. Hence, this optimization model would necessarily be the subject for a Ph.D. dissertation.

Conclusion: Option A, Optimal Work Duration on Site

Clearly, the latest literature review has shown that in order to develop an appropriately sophisticated optimization model, to account for the various parameters, which would allow for construction industry use of this model, considerably more time would be necessary at a higher degree of research (that is, Ph.D.) for this to be adopted as the second "Go Forward" case for this research. If this research were to proceed, the author would need to further evaluate, most likely by industry marketing survey, the perceived construction business' value of this option (an appetite to use this model as a project planning tool), given such a large gap in the literature for this option.

3.3.6.3 Workshop Panel Question No. 3

The author to proceed on Option B, Construction Site Waste Reduction, and investigate all possible scope for further detailed research for this textbook by further literature review, to open up available research themes.

After a specific review of construction site waste in over 50 academic journals, professional trade papers and government reports, and a subsequent compilation of a literature review information acquisition database to address this question, it was determined that there were four research methodologies that deserve further review for use as an effective tool, to assist in determining efficient construction waste mitigation processes. The identified void in the literature pertained to the need for research of front-end waste mitigation strategies on Australian construction sites.

I. The first technique was dynamic modelling to simulate the complicated construction site waste supply chain, commencing with off-site manufacture and fabrication of permanent and temporary materials and ending with construction material wastage being either placed in landfill, buried in regulatory approved locations on site or recycled as a new product for re-use. Dynamic modelling has many benefits for project stakeholders, including application for planning purposes to determine wastage type and volume prediction, which benefits contractors by planning and optimizing waste materials and assists the regulatory authority by aiding landfill footprint calculations. Once calibrated, dynamic modelling can also be used to check the actual wastage progressively on site versus model simulation to determine whether site material wastage is being adequately controlled (Hao et al. 2007, 2008, 2010).

II. The second research approach is a requirement to undertake a cost–benefit analysis, which would demonstrate exactly what commercial savings would result in a reduction in site construction waste for owner and contractor stakeholders as well as the community at large and the environment (Begum et al. 2006; Hwang and Zong 2011; Kozlovská and Spišáková 2013).

III. The third approach considers the development of an effective optimization model that would assist in determining the optimal processes to efficiently minimize the largest volume of construction waste on a project site (Yeomans 2004). The purpose of this optimization study would be to provide practitioners with a model that would assist in planning the optimal methodologies for the collection, allocation and disposal of physical construction waste.

IV. Finally, after carrying out a global literature review of available academic, regulatory and professional trade papers on front-end construction waste reduction, solicit assistance from senior Australian construction and engineering consultant industry executives to develop and undertake a survey on optimal front-end construction site waste reduction processes and address the void in the Australian academic literature.

3.4 "Go Forward" Case Recommendation

After consultation with peak industry group executives from the Australian Construction Industry Forum and the Master Builders Association, along with subsequent consultation with several peer review panellists and co-authors, it was decided
that for this textbook, a topic considering the front-end strategies on reducing waste
on Australian construction projects would provide the most effective and efficient
commercial, ecological and social benefits to the Australian engineering construction
industry stakeholders, the environment and the broader community.

There has been a significant amount of the literature written on downstream construction and demolition waste recycling over the past two decades, which aligns
with Australian government initiatives, and indeed a worldwide drive, to adopt sustainable construction waste processes, maximize recycling and, therefore, reduce
landfill footprints. By the end of the first decade of this century, significantly more
construction and demolition waste in Australia was recycled, rather than disposed in
landfill (Edge Environment 2012).

Notwithstanding this, in the opinion of the author, the positive international focus
on creating a "sustainable" world has been a *double-edged sword* with respect to
construction waste (United Nations 2015). Between the 1970s and 1990s, though
the academic literature in construction and demolition waste mitigation is scant,
there was a real root cause focus on reducing upstream/front-end waste, rather than
addressing how to handle the downstream waste product more effectively (Bossink
and Brouwers 1996; Skoyles and Hussey 1974; Spivey 1974). During this new century, international governments, including Australia, have generally pursued a recycling and resource recovery path to addressing the burgeoning increase of waste as
populations become more urbanized and wealthier (Chandler 2016). This literature
review has correspondingly shown that there has been little appetite to undertake
serious efforts to reduce the volume of construction waste at the source.

Refer to Fig. 3.3, "Construction Waste Management Hierarchy", adapted from
Faniran and Caban (1998), which lucidly demonstrates that the best methodology
for handling construction waste is to remove waste at the front end of the site materials
logistics chain, rather than downstream recycling/re-use recovery processes, which
not only cost money for transportation and/or production but produce accompanying
carbon emissions.

Accordingly, there is sufficient scope for further research to investigate the topic
of "*Front-End Strategies to Reduce Waste on Australian Construction Sites*" and fill
the void in the Australian academic literature.

Fig. 3.3 Construction waste
management hierarchy.
Adapted from Faniran and
Caban (1998)

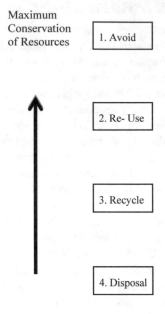

Maximum
Conservation
of Resources

**Construction Waste
Management Hierarchy:**

(1) Avoiding waste

(2) Re-using materials

(3) Recycling and reprocessing materials

(4) Waste disposal (if first three options are not possible)

3.4.1 Revised Textbook Title

Pursuant to the above, it was recommended that the revised title of the authors' book
was changed to: *"Effective Front-End Strategies to Reduce Waste on Australian
Construction Projects"*.

3.4.2 Revised Research Question

In view of this qualification to the "Go Forward" case for further research, a revised
"research question" was required at the conclusion of phase two. The revised research
question for this textbook is noted as follows: *"what effective and efficient front-*

Table 3.6 Revised research question

End of phase two research question construct

Research question criterion	Criterion description	Remarks
Interrogative word	What…?	The interrogative "what" explicitly presupposes a process
Phenomenon	Front-end waste reducing strategies	By research question inference, these processes shall be effective and efficient. Note textbook KPIs and case selection ensure optimal triple bottom line value
Context	Australian construction projects	–

By P. Rundle, 2017 (Adapted from Swinburne 2016)

end strategies are available to reduce waste on Australian construction projects?" Table 3.6 was utilized to objectively prepare this research question (Swinburne 2016).

3.4.3 Study Streams for Single Option "Go Forward" Case for Site Waste Reduction

There are several possible topics (for a single option covering the mitigation of construction waste at the front end of the waste cycle on Australian construction sites) to be further researched during the next stage of this textbook: phase three detailed research.

These topics are summarized below as:

I. Quantitative research to develop a dynamic model that simulates a project site's flow of construction waste for Australian projects. It was deemed by the author in consultation with his co-authors that this work is beyond the scope of a this book.

II. Quantitative research to develop an optimization model that improves construction site waste for Australian projects. It was deemed by the author, in consultation with his co-authors, that this work is beyond the scope of this research.

III. Carry out research to fill a void in the Australian academic literature to find possible practical solutions to reduce site construction waste by sourcing an appropriate seminal overseas paper and building on this knowledge locally. Similarly, the major causes of construction waste on Australian sites could also be evaluated.

IV. Develop an upper-level cost–benefit analysis (CBA) of sufficient detail to be used as a tool to: encourage the effective transfer of knowledge between this academic paper and practitioner use; foster change management processes by providing interested stakeholders with a commercial incentive to adopt these practices; and, because of inherent social and ecological benefits, have Australian governments promote these initiatives. In November 2017, after consultation with the SCU economics professor, it was agreed between the authors that that it was necessary to delete this CBA from this study scope because the nature of the work was deemed to be more suitable for Ph.D. work. Refer to Sect. 1.1, "Context", for further details.

3.4.4 Final Research Recommendation—One Option On-Site Waste with Two Study Streams

It was recommended that the author undertook detailed research on the reduction of construction site waste on Australian projects, using known strategies that minimize waste at the front end of the physical site material logistics chain. Accordingly, this book was re-titled "*Effective and Efficient Methodologies in the Australian Engineering Construction Industry—Front-End Strategies to Reduce Waste on Australian Construction Projects*". Furthermore, the "research question" was amended to read: "*what effective and efficient front-end strategies are available to reduce waste on Australian construction projects?*".

During the final stage of the literature review (in Chap. 5, "Research Methods"), after the development of a suitable research framework, a detailed global literature review will be carried out to identify the appropriate models and surveys for use in this research. The relevant local academic literature shall be studied in depth to clearly identify the gaps in the knowledge that this research shall build upon.

During the bi-weekly meeting, of 21 August 2017, the author's recommendations were formally agreed upon by the co-authors. On 22 November 2017, the cost–benefit analysis (CBA) was formally deleted from the scope of this textbook.

References

ABC. (2014). Global competitiveness, government regulation and productivity growth: Where does Australia rank? ABC News, 9 April.

ABS. (2013). *Survey of education and work* (Cat. no. 6227.0). Canberra: ABS.

ABS. (2015). *Labour force*. Canberra: Detailed Quarterly, May, ABS.

Ameh, O. J., & Daniel, E. L. (2013). 'Professionals' views on material wastage on construction sites. *Technology Management Construction, 5,* 747–757.

Australian Government Department of Employment (AGDE). (2015). *May 2015 industry outlook: Construction*. Canberra: Department of Employment, Australian Government.

Aykin, T. (1996). Optimal shift scheduling with multiple break windows. *Management Science, 42*(4), 591–602.

Bahn, S., & Barratt-Pugh, L. (2014). Health and safety legislation in Australia: Complexity for training remains. *International Journal of Training Research, 12*(1), 57–70.

Barney, J. B., & Hesterly, (2010). *VRIO framework. In strategic management and competitive advantage.* USA, Pearson.

Building and Construction Industry Organization (BCITO). (2015). *Older workers in the construction industry.* New Zealand: BCITO.

Begum, R. A., Siwar, C., Pereirea, J., & Jaafar, A. H. (2006). A benefit–cost analysis on the economic feasibility of construction waste minimisation: The case of Malaysia. *Resources, Conservation and Recycling, 48,* 86–98.

BHP Billiton Risk. (2007). *BHP Billiton energy coal hazard and risk procedures—rev 7, BHP Billiton project execution manual.* Australia: BHP Billiton.

Bossink, B. A. G., & Brouwers, H. J. H. (1996). Construction waste—quantification and source evaluation. *Journal of Construction Engineering and Management, 122*(1), 55–60.

Castanada, J. C. (2002). *Workers skills and receptiveness to operate under tier 2 construction management strategy* (Dissertation). University of Texas.

CIOB. (2015). Older workers crucial to curbing construction industry skills gap, but not a substitute for investing in training. *UK CIOB,* December. Retrieved June and July 2017 from http://www.ciob.org/media-centre/news/older-workers-crucial-curbing-construction-industry-skills-gap-not-substitute.

Chandler, D. (2016). *Construction waste is big business.* Construction News, 12 October. Retrieved May 2017–May 2018 from https://sourceable.net/construction-waste-is-big-business/.

Chase, E. (2009). *The brief origins of May day,* Retrieved April 2017 from https://www.iww.org/history/library/misc/origins_of_mayday.

Chowdry, R. M. (2014). Behaviour-based safety on construction sites: A case study. *Accident Analysis and Prevention, 70,* 14–23.

Dembe, A., Erickson, J. B., Delbos, R. G., & Banks, S. (2005). The impact of overtime and long work hours on occupational injuries and illnesses: New evidence from the United States. *Occupational and Environmental Medicine, 62*(9), 588–597.

Detsimas, N., Coffey, V., Sadiqi, Z., & Li, M. (2016). Workplace training and generic and technical skill development in the Australian construction industry. *Journal of Management Development, 35*(4), 486–504.

Dozzi, S. P., & Abourizk, S. M. (1993). *Productivity in construction.* National Research Council of Canada—Construction (NRCC—37001).

Edge Environment Pty. Ltd. (2012). *Construction and demolition waste guide for recycling and re-use across the supply chain.* Department of Sustainability, Environment, Water, Population and Communities. Retrieved May to July 2017 from https://www.environment.gov.au/system/files/resources/b0ac5ce4-4253-4d2b-b001-0becf84b52b8/files/case-studies.pdf.

European Union (EU). (2017). *Construction and demolition waste.* Retrieved December 2017 from https://www.interregeurope.eu/policylearning/news/1770/construction-and-demolition-waste/.

Faniran, O. O., & Caban, G. (1998). Minimizing waste on project construction sites. *Engineering Construction and Architectural Management, 17*(1), 57–72.

Fielden, S. L., Davidson, M. J., Gale, A. W., & Davey, C. L. (2000). Women in construction: The untapped resource. *Construction Management and Economics, 18*(1), 113–121.

Hao, J. L., Hills, M. J., & Huang, T. (2007). A simulation model using system dynamic method for construction and demolition waste management in Hong Kong. *Construction Innovation, 7*(1), 7–21.

Hao, J. L., Hill, M. J., & Shen, L. Y. (2008). Managing construction waste on-site through system dynamics modelling: the case of Hong Kong. *Engineering, Construction and Architectural Management, 15*(2), 103–113.

Hao, J. L., Tam, V. W. Y., Yuan, H. P., Wang, J. Y., & Li, J. R. (2010). Dynamic modeling and demolition waste management processes: An empirical study in Shenzhen, China. *Engineering, Construction and Architectural Management, 17*(5), 476–492.

Heaton, A. (2015). *Is Australia headed for a construction trade shortage?* Retrieved May 2017 from https://sourceable.net/is-australia-heading-for-a-construction-trade-shortage/.

Heaton, A. (2017). *Are older workers forgotten in construction safety?* Retrieved July 2017 from https://sourceable.net/older-workers-forgotten-construction-safety/.

Hickey J. (2015). *Landfilling construction waste in Australia.* Retrieved April to December 2017 from https://sourceable.net/landfilling-construction-waste-in-australia/.

Horner, R., & Talhourne, B. (1995). *Report on effects of accelerated working days and disruption of labour.* UK: Chartered Institute of Building.

Hwang, B., & Zong, B. (2011). Perception on benefits of construction waste management in the Singapore construction industry. *Engineering, Construction and Architectural Management, 18*(4), 394–406.

The Illinois Labor History Society (ILLHS). (2008). *The haymarket martyrs.* Chicago: ILLHS.

International Labour Organization (ILO). (2017). *About the ILO.* Retrieved April 17, 2017 from http://ilo.org/global/about-the-ilo/lang–en/index.htm.

Jaillon, L., Poon, C., & Chiang, Y. (2009). Quantifying the waste reduction potential of using prefabrication in building construction in Hong Kong. *Waste Management, 29,* 309–320.

Javelin, (2018). *Composition of construction and demolition waste.* Australia: Javelin Associates.

Keay, G. (2017). *Shutting the border: The impact of visa restrictions in Australia.* Retrieved July 2017 from http://constructive.net.au/latest-news/impact-of-visa-restrictions-in-australia/.

Kepner, C. H., & Tregoe, B. B. (2013). *The new rational manager.* Princeton, USA: Kepner Tregoe Inc., Publishing.

Knowles, R. (2002). Time is the enemy. *International Construction, 41*(8), 17.

Kozlovská, M., & Spisáková, M. (2013). Construction waste generation across construction project life-cycle. *Organization, Technology & Management in Construction, 5*(1), 687–695.

Langston, C. (2014). Construction efficiency: A tale of two countries. *Energy, Construction & Architectural Management, 21*(3), 320–325.

Lingard, H., & Lin, J. (2003). Career, family and work environment determinants of organizational commitment among women in the Australian construction industry. *Construction Management and Economics, 22*(4), 409–420.

Lingard, H., Zhang, R., Harley, J., Blismas, N., & Wakefield, R. (2014). *Health and safety culture* (pp. 1–124). RMIT University Centre for Construction Work Health & Safety Report.

Loosemore, M. (2015), *Work force diversity is key to construction innovation.* Retrieved July to August 2017 from https://newsroom.unsw.edu.au/news/art-architecture-design/workforcediversity-key-construction-innovation.

Lundberg, D., & Marshallsay, Z. (2007). *Older workers perspectives on training and retention of older workers: Support document—South Australian construction industry* (Research Report). South Australia: University of South Australia.

Marx, K. (Ed.). (1867). *Das capital (capital).* USA: Cosimo Incorporated.

Monuments Australia. (2017). *Eight hour day monument—Melbourne.* Retrieved May 01, 2017 from http://monumentaustralia.org.au/themes/culture.

Nagapan, S., Rahman, I., & Asmi, A. (2012). Factors contributing to physical and non-physical waste generation in construction. *International Journal of Advances in Applied Sciences, 1*(1), 1–10.

Organisation for Economic Co-operation and Development. (2017). *Economic surveys—Australia.* Retrieved July 2017 form https://www.oecd.org/eco/surveys/Australia-2017-OECD-economic-survey-overview.pdf.

Planet Ark. (2018). *How does Australia compare to the rest of the world?* Retrieved April 25, 2018 from https://recyclingweek.planetark.org/recycling-info/theworld.cfm.

Pugh, S. (1991). *Total design: Integrated methods for successful product engineering.* Boston: Addison Wesley Publishers.

Raheem, A., & Hinze, J. (2014). Disparity between safety standards—a global analysis. *Safety Science, 70,* 276–287.

Ritchie, M. (2016). State of waste 2016—current and future Australian trends. MRA Consulting Newsletter, 20 April. Retrieved March to April 2018 from https://blog.mraconsulting.com.au/2016/04/20/state-of-waste-2016-current-and-future-australian-trends/.

Safe Work Australia. (2012). *Australian work health and safety strategy 2012–2022.* Safe Work Australia. Retrieved April to June 2017 from https://www.safeworkaustralia.gov.au/about-us/australian-work-health-and-safety-strategy-2012-2022.

Skoyles, E., & Hussey, H. (1974). Wastage of materials. *Building, 21,* 91–94.

State of the Environment Overview. (SOE) (2016). *2016 State of the environment overview.* Retrieved May 2017 from https://www.environment.gov.au/science/soe.

Spivey, D. (1974). Construction solid waste. *Journal of the American Society of Civil Engineers, Construction Division, 100,* 501–506.

Srour, I., Haas, C., & Morton, D. (2006). Linear programming approach to optimize strategic investment in the construction workforce. *Journal of Construction Engineering and Management, 132*(11), 1138–1166.

State Library Victoria. (2017). *Workers' rights.* Retrieved April to May 2017 from http://ergo.slv.vic.gov.au/explore-history/fight-rights/workers-rights.

Swinburne University. (2016). *How to write a research question.* SU School of Electrical Engineering. Retrieved March to October 2017 from https://www.youtube.com/watch?v=lJS03FZj4K.

The Australian Industry Group. (AIG). (2015). *Australia's construction industry profile and outlook.* Retrieved April to June 2017 from https://www.aigroup.com.au/policy-and-research/economics/constructionoutlook/.

United Nations. (2015). *Sustainable development goals.* United Nations. Retrieved July to August 2017 from http://www.un.org/sustainabledevelopment/development-agenda.

University of New South Wales (UNSW). (2016). *Construction Industry: Demolishing gender structures.* UNSW Report.

Waste Management World. (2016). *State of the nation.* Waste Management World Periodical. Retrieved November 2017 to February 2018 from https://www.waste-management-world.com/au/report-state-of-waste-2016-current-and-future-trends.

World Economic Forum. (2014). *Global competitiveness report.* World Economic Forum. Retrieved June 2017 from www.weforum.org/docs/WEF_GlobalCompetitivenessReport_2014-15.pdf.

Yeomans, J. (2004). Improved policies for solid waste management in the municipality of Hamilton-Wentworth, Ontario. *Canadian Journal of Administrative Sciences, 21*(4), 376–382.

Yi, W., & Chan, A. (2014). Optimal work pattern for construction workers in hot weather: A case study in Hong Kong. *Journal of Computing in Civil Engineering, 29*(5), 05014009.

Zero Waste South Australia. (2013). Annual Report 2012–2013. Government of South Australia. Retrieved May to June 2017 from http://www.zerowaste.sa.gov.au/upload/resource-centre/publications/corporate/3ZWSA%20Annual%20report%202013%20DE_02.pdf.

Chapter 4
Core Study Objectives

It is appropriate at the completion of phase two of this textbook, that the core research objectives be described. They are summarized below.

4.1 Aims

The purpose of this research is to gather information on the most effective and efficient methods for the reduction of waste on construction sites in the Australian environment, using both existing international research and data from members of the local construction industry. The author expects that the identified void in the academic literature shall be filled, in part, and that this textbook will pave the way to further research on this topic. Further, through thoughtful knowledge transfer, post-textbook completion, the ultimate aim of the research and the triple bottom line value shall be achieved and enjoyed by the contractor/builder/client stakeholders, the community at large and the environment, through the implementation of these more effective and efficient construction site waste mitigation strategies.

4.2 Contribution to the Engineering Construction Industry

This research has been conducted to provide members of the engineering construction industry with information regarding the types of site waste reduction practices that are currently being implemented and the reasons for their adoption on construction sites. The research was also undertaken to assist in effective decision-making processes regarding the implementation and evaluation of efficient construction waste reduction methodologies on field projects. A void in the literature has uncovered a very real need for this research to be carried out in Australia, similar to a recent research project

© Springer Nature Switzerland AG 2019
P. G. Rundle et al., *Effective Front-End Strategies to Reduce Waste on Construction Projects*, https://doi.org/10.1007/978-3-030-12399-4_4

commissioned in the USA (Yates 2013), and to carry out further academic research to build upon the knowledge acquisition database created by a seminal Australian work by Faniran and Caban (1998), by using their adapted model.

4.3   Broader Community Benefit

This section has been further expanded via detailed literature review from the information on site waste minimization, provided earlier in this book, in Chap. 3, ""Short-List" of Four Methodologies", which was ultimately used for briefing purposes during the peer review "Go Forward" case decision analysis workshop.

Depending on the type, size and location of construction projects in Australia, anything between eight and 30 per cent of construction materials are wasted and disposed of as landfill or recycled in Australia and there are several proprietaries off the shelf estimating software packages, which have the capability to estimate construction material waste quantities, which vary between material product and constructed element (Core Logic 2016; Doab 2018). A 2017 Scottish Government report indicated that an average of 13 per cent of construction material was wasted on projects in that country and that a limited approach in waste mitigation would result in a one per cent saving to a project's bottom line (Zero Waste Scotland 2017). A University of Exeter report on construction waste minimization suggests that by adopting a set of ten strategies, project capital costs can be reduced by two per cent; they have integrated this process in their capital work programme (Vowles 2011). These ten strategies are quoted as follows:

> The proposed actions are therefore listed in chronological order as the project progresses:
>
> (a) Provide regular training for EDS staff to update on issues with construction waste and means to minimise it and embed this within the EDS ISO9001 QA [Quality Assurance] system.
>
> (b) Include within the project brief and the tender documents for the consultant's services and construction works, the University's aim and requirements to minimise the environmental impact caused by any construction works.
>
> (c) The project team will consider means to minimise construction waste from the inception to the completion of the project, through the initial brief, design process, materials selection, construction techniques and operational methods.
>
> Examples of specific requirements include:
>
> (a) Inclusion within the Scope of Services for the appointment of designers, a requirement to comply with the 'Designing out waste: a design team guide for buildings' guideline. This guide encourages designers to design out waste by signing up to the five key principles design teams can use during the design process to reduce waste, namely to:
>
> - Design for re-use and recovery
> - Design for off-site construction
> - Design for materials optimization
> - Design for waste efficient procurement

- Design for deconstruction and flexibility
- Production of whole life cycle costs for key structural and services elements of construction projects during the design process
- Inclusion within the EDS ISO9001 QA system, as part of the briefing process for refurbishment/alterations projects, a requirement for the Project Manager/Designer to question the need to replace fittings and fixtures in buildings, unless it is economically viable to retain them (Vowles 2011, pp. 1–3).

Waste Management World's 2016 "*State of the Nation*" reported that:

construction and demolition (C&D) waste (typically timber, concrete, plastics, wood, metals, cardboard, asphalt and mixed site debris such as soil and rocks) comprises approximately 40 per cent of Australia's total waste generation (Waste Management World 2016, n.p.).

A report commissioned by the Federal Government shows that C&D waste accounted for 31 per cent of landfill waste in Australia (Hyder Consulting 2011b). Chandler (2016) states that the Australian construction industry is the largest "offender" of any industry, accounting for a full third of all waste sent to landfill.

Because of the large number of unregistered active landfills in Australia, many private facilities do not have weighbridges and it is difficult to collate accurate waste data in Australia. Ritchie (2016, n.p.) argues that:

[w]hile the total number of active landfills in Australia is unknown, Commonwealth Government data indicates there are at least 600 mid to large sites, while there could be as many as 2000 unregistered and unregulated landfills. The fact that, as a nation, we are unsure of the exact number of landfills in Australia, requires immediate review. Small, unlined landfills can still have localized (environmentally detrimental) impacts.

Any decrease in construction waste brought about by adopting front-end site waste minimization will have major positive ecological benefits to Australia by reducing unregulated landfill footprints, many of these landfills, according to Richie (2016), being potentially contaminated.

There shall always be a requirement for a landfill levy, based on market viability of recycling. In 2013, the Queensland Government brought in a A$35 per tonne industry level that created an immediate increase of recycling rates. However, when the levy was removed, 18 months later with a government change, recycling rates plummeted by 15 per cent (Waste Management World 2016). This example of what "the market dictates" illustrates that pursuing front-end waste reduction strategies, which could actually reduce project costs, would find ready acceptance in the engineering construction industry.

It has been estimated that construction waste has amounted to A$2.8 billion, annually, for residential building construction in Australia, not including high-rise apartment dwellings (Chandler 2016). Conservatively, the cost of construction waste on major heavy construction projects in Australia totalled 6.5 per cent of total job cost and significantly more for smaller projects (Doab 2018). Jain (2012) carried out a study on the economic aspects of construction waste by assuming a waste rate of 13 per cent when reviewing a number of construction sectors. They recommended an average 50 per cent material cost as part of the total project cost across all sectors

(Jain 2012). This amounted to an additional 6.5 per cent waste impost onto Indian projects (Jain 2012).

The Master Builders Association of Victoria (MBAV) argues that the cost of construction site waste adds 10 per cent to the cost of a building construction in Australia (MBAV 2004). The high-cost impost on capital work projects due to physical material construction wastage, ranging from residential housing to mega-multi-billion dollar resource projects, would see massive savings to both the private and the public sector building and construction industry if the two per cent savings that are mooted by following the ten front-end strategies recommended by the University of Exeter (Vowles 2011) were adopted. Josephson and Saukoriipi (2007), on their review of non-physical and physical aspects of project construction waste in Sweden, found that the cost of physical material waste (excluding scope changes) came to be between 6 and 11 per cent of a project case, depending on which of the several case studies were evaluated.

The cost of all construction in Australia, in 2016, was A$218 billion, which comprised: residential construction (A$96 billion); non-residential construction (A$37 billion); and engineering construction (A$85 billion) (Australian Construction Industry Forum (ACIF) 2017). A savings to the nation's engineering construction industry of two per cent or A$4.36 billion is possible.

Chandler (2016) summarizes that the construction industry is the globe's major consumer of raw materials, but that around only one-third of construction and demolition waste is recycled or re-used, as per a recent World Economic Forum report. Australia is one of the ten highest producers of solid waste in the OECD (Chandler 2016). Similar to Australia, solid waste in the USA typically comprises around 40 per cent of construction and demolition waste per annum (Chandler 2016).

The total savings attributable to a reduction in construction material site wastage shall include, but not necessarily be limited to:

I. Carbon footprint reductions via manufacture, transportation and disposal/recycling of significantly less construction materials;
II. Savings in energy and water via significantly less construction material manufacture and recycling;
III. Ecological benefits to the groundwater and aquifer systems, due to landfill reduction;
IV. Savings in the costs of transportation, procurement and storage, and landfill;
V. Environmental benefits to the broader community via pollution reduction;
VI. Less impact on environmental footprint via reduction of landfill size and a reduction of unregulated and possibly contaminated landfill sites; and
VII. Savings in rare and valuable irreplaceable raw resources, such as lime, timber, sand, clay, rock, gypsum, coking coal for steel manufacture, titanium, copper, bauxite and iron ore.

Furthermore, a reduction in the above-noted estimated cost of A$2.8 billion per annum for residential construction waste could greatly assist home buyers via a reduction in construction costs. Contractor, builder and client stakeholders could all benefit from project capital cost reduction of the estimated A$4.36 billion per annum due to front-end waste reduction, calculated, above, in this Sect. 4.3 (ACIF 2017).

4.4 Research Impact

This book shall endeavour to capture, organize, combine and share the knowledge base comprising the literature review data and research data, to provide value and capabilities in new and distinctive ways for interested industry stakeholders (clients, contractors, builders and engineering consultants). This body of research knowledge could enhance an organization's competitiveness.

However, in practice, transferring knowledge within the construction industry sector is historically difficult to achieve (Argote et al. 2000; Fellows et al. 2009), and in order to maximize the impact of this textbook, research of the optimal communication vehicles for knowledge flow shall be appraised. At the mid-point of this research, it has been determined that the optimal means to maximize research impact is the dissemination of this textbook through peak industry bodies: Engineers Australia, Consult Australia, Australian Constructors Association, Master Builders Association and Australian Construction Industry Forum. Awareness of this research will also be increased via publications in a suitable academic journal, such as the "*American Society of Civil Engineers*", and trade papers such as the Australian edition of "*Construction News*". Both the Australian Constructors Association and Australian Construction Industry Forum, when consulted, indicated a preference for further research on effective and efficient methods to reduce site construction waste.

References

Argote, L., Ingram, P., Levine, J., & Moreland, R. (2000). Knowledge transfer in organizations. *Organizational Behaviour and Human Decision Processes, 82*(1), 1–8.

Australian Construction Industry Forum (ACIF). (2017). *Latest forecast*, Retrieved March to April 2017, from https://www.acif.com.au/forecasts/summary.

Chandler, D. (2016). Construction waste is big business. Construction News. 12 October, Retrieved May 2017–May 2018, from https://sourceable.net/construction-waste-is-big-business/.

Core Logic. (2016). *Cordell cost guide*. https://www.corelogic.com.au/products/cordellcostguides, viewed October–December, 2017.

Doab, M. (2018). Average cost of construction site waste on major Australian projects. In *Doab estimation enterprises proprietary*. D'Est Estimating Model.

Faniran, O. O., & Caban, G. (1998). Minimizing waste on project construction sites. *Engineering Construction and Architectural Management, 17*(1), 57–72.

Fellows, R., T., Langford, D., Newcomb, R., & Urry, S. (2009). *Construction management in practice*. UK: Blackwell Science.

Hyder Consulting. (2011b). *Department of sustainability, environment, water, population and communities—waste and recycling in Australia 2011 incorporating a revised method for compiling waste and recycling data*. Australian Commonwealth Government. Retrieved May to July

2017, http://www.wmaa.asn.au/event-documents/2012skm/swp/m2/1.Waste-and-Recycling-in-Australia-Hyder-2011.pdf.

Jain, M. (2012). Economic aspects of construction waste materials in terms of cost savings—a case of Indian construction industry. *International Journal of Scientific Research Publications, 2*(10), 1–7.

Josephson, P., & Saukoriipi, L. (2007). *Waste in construction projects: Call for new approach.* Sweden: Chalmers University of Technology Publication.

Master Builders Association of Victoria (MBAV). (2004). *The resource efficient builder–A simple guide to reducing waste.* MBAV Guideline Publication.

Ritchie, M. (2016). State of waste 2016—current and future Australian trends. MRA Consulting Newsletter. 20 April, Retrieved March to April 2018, from https://blog.mraconsulting.com.au/2016/04/20/state-of-waste-2016-current-and-future-australian-trends/.

Vowles, F. (2011). *Construction strategy.* University of Exeter. Retrieved November to December, 2017 from https://www.exeter.ac.uk/media/universityofexeter/campusservices/sustainability/pdf/construction-strategy.pdf.

Waste Management World. (2016). *State of the nation.* Waste Management World Periodical. Retrieved November 2017 to February 2018, from https://www.waste-management-world.com/au/report-state-of-waste-2016-current-and-future-trends.

Yates, J. (2013). Sustainable methods for waste minimisation in construction. *Construction Innovation, 13*(3), 281–301.

Zero Waste Scotland. (2017). *The cost of construction waste.* Government of Scotland. Retrieved February to June, 2017 from http://www.resourceefficientscotland.com/content/cost-construction-waste.

Chapter 5
Research Methods

5.1 Research Phase Literature Review

The author was fortunate in having a SCU professor from a technical discipline with a qualitative research background to act as a mentor through the qualitative research framework and the qualitative methods' data synthesis model development stage of this research. This involved extensive academic literature sourcing to derive the research framework and related methodologies. The author also wished to undertake one final interrogation of the available local Australian academic literature to verify any "available" scope in the missing void literature that could be exploited for further research. This literature was discovered by backtracking through Australian professional practice and trade paper references to academic journal articles.

5.1.1 Site Waste Sources and Minimization

There has been considerable academic, professional and government literature produced in the past 20 years regarding sustainable construction waste treatment methodologies (Ibrahim 2016; Lu et al. 2017; Poulikakos et al. 2017; Rodrigues et al. 2013; Song et al. 2017; Wahi et al. 2016; Yates 2008). The majority of international researched literature indicates that the primary emphasis of these papers is on the downstream treatment of construction waste as a recycled by-product, thus reducing the landfill footprint. Indeed, much of the academic literature on Australian construction waste surrounds the recycling aspects (Tam and Tam 2006; Tam 2009; Treloar et al. 2003; Hiete et al. 2011).

There has been some recent Australian academic research on front-end construction site waste, considering such factors as life-cycle planning and human-centred variables (Ezo et al. 2017; Udawatta et al. 2015; Wu et al. 2017). None of these papers have used the seminal works, traceable back to Spivey (1974), through to Bossink and Brower (1996) and the seminal Faniran and Caban (1998) sources of site waste

© Springer Nature Switzerland AG 2019
P. G. Rundle et al., *Effective Front-End Strategies to Reduce Waste on Construction Projects*, https://doi.org/10.1007/978-3-030-12399-4_5

model. This has limited the potential of these papers to compare their research with previous papers following seminal approached—particularly, for research data reliability and validity purposes. The adoption by the author of the Faniran and Caban (1998) model and the Yates (2013) model for qualitative research surveys has allowed a broad-based comparison with other academic literature. In addition, the sample limitations, outlined in the Udawatta et al. (2015) research conclusions, were considered in the author's research. The project sample was increased to include all major sectors of the Australian engineering construction industry, including infrastructure, mining and process, and residential/non-residential building.

Yates (2013, p. 282) posits that the relevant literature on construction site waste concentrates on the building industry rather than on heavy infrastructure projects and that both segments of the construction business need to be analysed. Yates (2013, pp. 291–292) contends that front-end waste minimization, on-site, is also a critical element in sustainable construction practices that must be evaluated.

In 1994, Task Group 6 of the Conseil International du Bâtiment (CIB 2017), an international construction research networking organization, defined the goal of sustainable construction as "creating and operating a healthy built environment based on resource efficiency and ecological design" (CIB 2017) and introduced seven principles of sustainable construction:

  I.   Reduce resource consumption.
 II.   Re-use resources.
III.   Use recyclable resources.
 IV.   Protect nature.
  V.   Eliminate toxins.
 VI.   Apply life-cycle costing.
VII.   Focus on quality (CIB 2017; Kibert 2008, p. 6).

The Conseil International du Bâtiment anglicized its name, in 1998, to the International Council for Research and Innovation in Building and Construction and its member representative in Australia is RMIT.

It has become apparent, during this current literature review, that it will be necessary to also introduce an extra survey questionnaire into the study of Australian project site waste minimization strategies and that this new topic should be further researched for identifying what the main sources of site waste are. This information will need to be researched so that the data on waste reduction methodologies can be matched against the site waste sources.

Nagapan et al. (2012) provided a qualitative approach for determining potential sources of construction waste, while a 2002 Griffith University conference paper from Sugiharto et al. (2002) adopted a quantitative approach for determining construction waste sources. However, a seminal Australian paper by Faniran and Caban (1998) was presented as a possible, often cited, model for use in this research on sources of construction waste. In fact, Nagapan et al. (2012) used an adapted Faniran and Caban (1998) sources of waste identification model in their Malaysian research.

Construction waste can be classified into physical and non-physical groups (Alarcon 1994). Physical construction waste is defined as waste which arises from con-

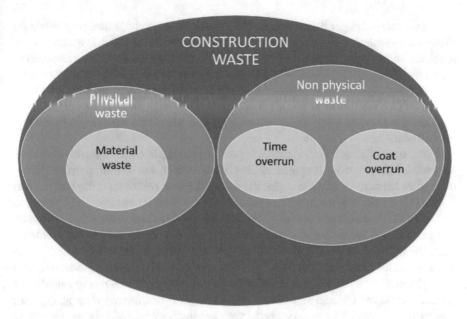

Fig. 5.1 Classification of construction waste (adapted from Nagapan et al. 2012)

struction, renovation and demolition activities, including land excavation or formation and civil/building construction (Alarcon 1997; Ferguson 1995; Formoso et al. 1999; Yates 2013). The physical waste product comprises inert temporary and/or construction material, including, but not limited to bricks, concrete debris, timber offcuts, tiles, glass, vegetation waste from site clearing, steel offcuts, bricks and blocks, plasterboard, plastic wrapping and pallets (Alarcon 1997; Ferguson 1995; Formoso et al. 1999; Yates 2013). Non-physical construction waste is generally created by construction process issues and can be defined as related to such matters as delays resulting from activities such as repairs, waiting time and delays causing project cost and schedule overruns, along with equipment downtime and the unnecessary movement of workers (Hao et al. 2008; Nagapan et al. 2012; Nazech 2008).

For the purposes of this book, only physically constructed non-permanent and permanent materials, wasted through the project supply chain, shall be included in this research. A study of non-physical waste created, in the most part, by project management processes would be an entirely different study. Permanent materials are defined as materials that are used in the built works, such as reinforced concrete; temporary materials are those such as concrete formwork/falsework materials, which are used in construction methods, but do not form part of the permanent element (Antill and Ryan 1979). Refer to Fig. 5.1, "Classification of Construction Waste", which provides a diagrammatic view of physical and non-physical waste.

After an extensive literature review of local academic journal articles; professional papers; trade papers; and government information sources pertaining to construction waste, a possible void in the literature was identified. There was a seminal Australian academic journal paper discovered in this research, entitled *"Minimizing Waste on Project Construction Sites"*, written by Faniran and Caban (1998), which used a research framework that could be traced back to the seminal paper on construction waste by Spivey (1974). Faniran and Caban developed their sources of waste model from another seminal work by Bossink and Brouwers (1996), adapted by Faniran and Caban (1998) in their research. This paper has used some solid research of Australian construction projects using Australian builders and contractor's participants, but this was 20 years ago.

Faniran and Caban's (1998) model is preferred to the later 2012 waste minimization model prepared by Nagapan et al. (2012) (which the author sourced during the earlier literature study phase) for the following four reasons: (i) the Faniran and Caban (1998) model is an Australian model adapted for this country; (ii) Faniran and Caban (1998) have clearly described their data collection and data analysis research framework, and the author was unable to obtain a pro forma survey questionnaire from the Nagapan et al. (2012) study; (iii) the Australian model (Faniran and Caban 1998) includes more classifications than Nagapan et al.'s Malaysian model; and (iv) Nagapan et al.'s paper was not peer-reviewed and appeared in a newly established journal, while the Faniran and Caban (1998) paper is considered a seminal paper. Nagapan et al. (2012) have undertaken a very exhaustive worldwide literature review and have added valuable work on this scantly researched topic.

The author eventually found two other Australian papers on construction waste by Sugiharto et al. (2002) and, more recently, Udawatta et al. (2015). However, neither of these papers used the Faniran and Caban (1998) approach that was traceable back to Spivey (1974) and Bossink and Brouwer (1996), which the author was keen on using to assist in data validation, internationally and in Australia. Nevertheless, both Udawatta et al. (2015) and Sugiharto et al. (2002) provided important research that was subsequently used by the author for data validity purposes.

On the matter of possible strategies to minimize construction site waste, the author opted to use the Yates (2013) paper because this work captured similar construction works that are occurring in Australia at present (large commercial building projects as well as major infrastructure construction). The Yates (2013) findings will serve as a benchmark to this research. Furthermore, Yates (2013) has thoroughly described their data collection and data analysis approaches and has provided details on the survey questionnaire that the author has adapted for use in this study by using the waste minimization strategy questionnaire and not the waste recycling component.

The Sugiharto et al. (2002) conference paper uses a qualitative method's approach that considers non-physical, as well as physical waste in its 2002 evaluation of Australian construction waste, which they define as "non-value adding" to the Australian construction industry, in line with a Lean Construction approach as outlined in their paper adapted from Koskela (2000). The key findings of the paper were that "lack of design" and "waiting for instructions" were the two major sources of waste in their research; both of these are non-physical waste resource productivity issues (Sugiharto et al. 2002). The reader's attention is drawn to Sect. 2.3.5, ""Long-List" Decision Analysis to Determine "Short-List"", of this textbook and more particularly to the fact that the independent peer review panel rated the Lean Construction process as the least value providing option in this book. Furthermore, none of the 10 panellists elected for this option to proceed. Subsequent literature review to validate this result found several papers, for example, Green and May (2003), which discredited this Lean Construction approach as being too "mechanistic".

Sugiharto at al. (2002) had 50 respondents who completed their survey which is in line with the respondent sample size of 53. However, the participant sample in that study comprised of contractor site managers, construction managers and project managers only. This research was undertaken 16 years prior to this textbook (Sugiharto et al. 2002).

As the author study is limited to physical waste only, the study research approach is qualitative. There are concerns with the Lean Construction approach used by Sugiharto et al. (2002), both from the 10 independent peer panel practitioners and the academia via Green and May (2003). The Sugiharto et al. (2002) paper was not adopted as a framework in this research, even though there are valuable findings in this well-presented work that the author evaluated for correlation with this textbook where possible.

A literature review by the author on the composition of physical construction waste material elements (e.g. concrete, steel, timber, glass and plasterboard) found in landfill disposal in an Austrian-based trade paper (Waste Management World 2016), alluded to an Australian consulting newsletter on waste (Ritchie 2016). The newsletter referenced an Australian paper on construction waste by Udawatta et al. (2015).

The Udawatta et al. (2015) study adopted a mixed methods research approach, using a statistical factor analysis method for data synthesis. The research received 104 survey responses from the survey sample cohort, all comprising of non-residential building project managers, who were all members of both the Australian Institute of Building and the Australian Institute of Project Management. The study focus was on the elimination, reduction or recovery treatment of construction waste, not sources of waste. The Udawatta et al. (2015) paper has provided several thoughtful recommendations, advancing Australian building site waste minimization practices, developing its survey questionnaires from a literature review and evaluating the human factors that had the potential to contribute to site waste reduction. In the conclusion of the Udawatta et al. (2015, p. 82) paper, they state that:

> [t]his study focused on ways of improving WM practices in non-residential building projects and the questionnaire was distributed only to project managers. Further research opportunities exist to explore how these solutions can be applied in different project types with different procurement methods and other stakeholder groups.

The author's approach in this study has built upon Udawatta et al.'s research by opening up the survey participant sample, as they suggest in their research conclusion, to encompass the whole range of residential and non-residential building in Australia, including high-rise work. Australian heavy construction, including resources, process and infrastructure construction, is also considered, as are Australian engineering construction firms working overseas. In addition, the respondent demographics show that the survey respondents included: engineering construction company directors, including a chairperson and chief executive officer (CEO); private and public sector capital works clients; project managers; structural general superintendents; high-rise site managers; estimators; project environmental scientists; and landfill and recycling managers.

This book has replicated the seminal work of Faniran and Caban (1998) on sources of waste, which Udawatta et al. (2015) did not consider in their research. The author carried out a literature review to discover an appropriate current academic paper that had surveyed senior engineering construction executives, in both the heavy construction and building sectors, which would lend itself to thematic analysis to enable that paper's research approach to match strategies to the corresponding sources of waste research. The Yates (2013) paper from the USA was that paper, with the research being funded by America's peak construction body and the paper being published by the American Society of Civil Engineers.

Another advantage of this approach was that the use of both the Faniran and Caban (1998) model and the Yates (2013) model in this research has allowed ready benchmarking with other academic research for validation purposes, for example, Nagapan et al. (2012). This research also uniquely considers both the sources of construction waste and the possible site waste minimization strategies.

5.2   Research Sub-question

During the formulation of this qualitative research methods section of the textbook, and also while undertaking further literature review as summarized in Sect. 5.1, "Research Phase Literature Review", it became apparent that it would be necessary to also undertake a survey of respondents to discover their views on the sources of site construction waste to match the recommended waste reduction strategies. According to the qualitative method pragmatic framework approach, a research sub-question would need to be developed to focus the research on physical site waste sources. This sub-question is complementary to the primary research question, which is: "*what strategies are available to reduce waste on Australian construction projects?*" Using the Swinburne University (2016) engineering faculty approach to research question development, the research sub-question to address further detailed qualitative research by survey questionnaire is: "*what are the major sources of waste on Australian construction sites?*"

5.3   Preferred Option Detailed Research Phase

In Sect. 5.3, the preferred option, "*Front-End Strategies to Reduce Waste on Australian Construction Projects*", shall be researched in depth, and the research question, "*what effective and efficient front-end strategies are available to reduce waste on Australian construction project?*", will be answered.

During this detailed research stage of the study, relevant overseas academic papers were sourced that could be adapted for use on this research. As an example, previously published questionnaires could be utilized in acquiring survey data from practitioner proponents. The research involving peak industry executive participants will validate or reject the acquisition knowledge base developed via a literature review. It is a criterion of this research that the findings shall be of a practical use to engineering practitioners.

It was recommended at this stage to develop an upper-level cost–benefit model, using existing available data, to provide a broad basis to allow the engagement of engineering construction practitioners with the academic concepts covered in this book, to facilitate the transfer of knowledge. However, acting upon a subsequent recommendation by a SCU economics professor who reviewed this cost–benefit analysis scope, and at the recommendation of the SCU confirmation panel, this work was deemed to be too detailed and was deleted from the scope of this research.

It shall be necessary to carry out a targeted literature review specifically for the extra identified data necessary for this research, such as subsequent extra information, which allows succinct and unambiguous definitions of key terms such as "construction waste". For instance, this includes research on existing academic literature on the construction waste supply chain to assist in the subsequent analysis of possible solutions. Appropriate overseas academic papers will be sought by further literature review that can be adapted for use in this study.

After the research methodology is developed, the tools for this research, such as the appropriate survey software, the qualitative survey questionnaires pro forma, the Likert scale quantitative survey pro forma, the cost–benefit model and the constant and variable constraints, shall be selected and/or developed. Potential survey participants shall be sourced and surveys distributed. Data shall be collated and analysed, findings shall be tabled, and conclusions drawn, along with recommendations for further academic research. However, the first phase in this research stage needs to be the preparation of an appropriate qualitative methods approach for conducting the research, as described in Sect. 5.4.

5.4 Qualitative Approach and Research Paradigm

The initial step in developing the research framework shall be to determine which research methodology, the quantitative or the qualitative approach would be the most appropriate methodology with which to conduct this research. A possible mixed methods option is also addressed later in this chapter.

Table 5.1 demonstrates the dichotomy between qualitative and quantitative methods of conducting academic research. A review of the academic literature surrounding the research question and of Table 5.1 clearly indicates that a qualitative approach is the appropriate methodology. The reason for this is that the detailed research effort to source effective and efficient front-end strategies to reduce waste on Australian

Table 5.1 Research method's dichotomy

Methodological dichotomy	
Qualitative research attributes—the "classic" post-positivist paradigm	Quantitative research attributes—the "classic" positivist paradigm
Subjective reality	Objective reality
Meanings	Causal
Human intentions	Detached
Personally involved	Samples/populations
Study cases	Contrived
Actors in natural settings	Variables
Verbal and pictorial data	Numerical
Generalise case findings	Statistical
	Impersonal

By Rundle (2017) (adapted from Cameron 2015)

construction sites will involve data collation to poll participant's subjective advice via questionnaire surveys, rather than an empirical study to mathematically address the research question using a quantitative approach.

Qualitative research methodology is typically inductive, emerging, and shaped by the researcher's experience in collecting and analysing the data. The logic that the qualitative researcher follows is inductive, from the ground up, rather than handed down entirely from a theory. On occasion, such as in this book, the research question can alter in the middle of the study to better address the types of questions required to resolve the research problem. During the data analysis, the researcher follows a path of analysing the data to develop an increasingly detailed knowledge of the topic being studied (Creswell 2013).

The next step to develop a research framework is to undertake a critical review of the following interpretive frameworks that it was believed might be applicable for qualitative research for this textbook:

 I. Post-positivism (Morse 1994);
 II. Social constructivism (Mertens 2009);
III. Post-modern perspectives (Bloland 1995); and
IV. Pragmatism (Tashakkori and Teddlie 2003).

Lincoln and Guba (1985) posit that combining qualitative and quantitative methods into one study is incompatible, while Patton (1990) states that both these methods can be combined. The author's literature research on the appropriate method, qualitative or quantitative, disclosed academic literature on the long-term 1970s' argument in academia between supporters of either a quantitative or qualitative research method, entitled the "paradigm wars" of the 1970s (Oakley 1999). This was basically caused by new and at times radical critiques of scientific research and a dislike of quantitative methods as a vehicle for collecting social science data (Oakley 1999).

It is the intention of the author to select an interpretive framework that is "fit for purpose", to best drive the qualitative approach forward with commensurate academic rigour to address the research question, yet provide sufficient latitude to allow a focus on practical outcomes, rather than embark on protracted discourse on the relative merit of varying philosophies and methods, or otherwise.

There would be a degree of unconscious bias in the selection of the appropriate paradigm to undertake this research, due to the author's personality traits and profession. However, a "self-awareness" session was carried out to limit the impact of unconscious bias on this selection (Gudmundsson and Lechner 2013).

Pragmatism was selected as the most appropriate interpretive framework to take this research forward using a qualitative methodology. The nature of this research, with a fundamental aim to benefit engineering practitioners and the engineering construction industry, and an abiding interest in finding practical solutions to problems, means that a pragmatic approach, focusing on research outcomes to directly address the research question, is most suitable. Inquiry actions, situations and consequences are more relevant than antecedent conditions, as per post-positivism (Patton 1990).

A pragmatic interpretive framework has a number of persuasive features that makes it the optimal paradigm for this research. Note these features, summarized as follows:

I. A pragmatic framework focuses on the problem being studied and addressing the research question, rather than a methods' typology focus (Rossman and Wilson 1985).
II. Pragmatism remains uncommitted to any single philosophy (Cherryholmes 1992).
III. Individual researchers have freedom of choice to choose methods, techniques and research procedures that shall meet their requirements (Creswell and Poth 2017).
IV. Using a pragmatic approach, the research in practice adopts an overarching international view for using multiple methods of data collection to best answer the research question. It shall also utilize multiple sources of data collection, so as to focus on the practical implications of the research and will foster a requirement to carry out research that best answers the research problem (Rorty 1966; Tashakkori and Teddlie 2003).
V. A pragmatic approach can be suitable for qualitative and quantitative approaches (Creswell and Poth 2017; Cameron 2015). Pragmatism is a particularly appropriate paradigm for use with a mixed method approach (Creswell et al. 2003). Pragmatism is a paradigm that is practical and applied, which supports the use of both qualitative methods and quantitative methods in the same study, rejecting incompatibility (Maxcy 2003).

The qualitative research design is phenomenological, with the purpose of this design being to reduce individual experiences into a common experience or theme. Creswell's (2013) view is the pragmatic paradigm, which focuses on what works to attain solutions, rather than on a particular method.

5.5   Framework for a Pragmatic Approach

The commensurate academic rigour that must be achieved in research to attain a quality product can be attained by using a suitable pragmatic approach framework (O'Brien et al. 2014). O'Brien et al. (2014) argue that it will be necessary to:

  I. Identify recurrent patterns in the data;
 II. Attend to the language used in the social context;
III. Attend to what is "normal"; and
 IV. Put into the context of the entire interaction.

As proffered by Seale (1999), research is in large part a craft learned through personal experience of doing research and from an appreciation of what is considered good in other research. "Quality … does not depend on unthinking adherence to rules of method … but exposure to methodological debates can help loosen thoughts that are stuck" (Seale 1999, p. 475).

Table 5.2, "Framework for a Pragmatic Research Approach", was developed by Pope et al. (2000) from British research standards for a surgical process study in a British hospital. The paper indicates that a study of this nature would usually be undertaken using a quantitative research approach, rather than the qualitative pragmatic approach adopted for the research. The author used this pragmatic approach framework because this research was often undertaken using a quantitative approach, such as that undertaken by Udawatta et al. (2015). The Pope et al. (2000) framework was an ideal approach for these site waste studies allowing flexibility of approach, including thematic analysis, yet yielding a rigorous research method regime.

5.5.1   Data Collection

This study employed a qualitative methods approach using the pragmatic paradigm to develop a framework for research. The data was analysed using a thematic content analysis approach to answer the research question. Methodological triangulation, using qualitative data derived from this research and data used for benchmarking purposes from seminal academic papers, using the same survey materials, was adopted to strengthen the validity of the data.

The survey questionnaires were prepared on proprietary Qualtrics software and delivered to the survey sample members, divided into four cohorts, using the Internet as the communication's vehicle to complete the study questionnaires. The first cohort of proponents was all members of one of Australia's three main peak engineering construction groups, the Australian Construction Industry Forum (ACIF). The ACIF managing director agreed to directly transmit the author's Qualtrics survey questionnaire to its 12 members on behalf of the author. The Master Builders Association (MBA) was contacted in Sydney without response; however, the Northern New South

Table 5.2 Framework for a pragmatic qualitative research approach

Framework for a pragmatic approach			
Standards	Terms associated with qualitative research	Function	Remarks
Framing		Show the research has purpose in the context	Make it clear when "framing" what is already known and what this study builds on. What is the "something" that has not been described before
Reliability	Procedural trustworthiness	Ensure respect for procedures	Justify the size and nature of the sample. The data must properly address the research question
Validity	Trustworthiness of the findings	Ensure respect for findings	Ensure analysis is realistic—use techniques associated with grounded theory (i.e. thematic analysis using open coding) Ensure findings are coherent with categories defined and mutually exclusive Do the findings show theoretical validity by developing theory? Do findings show catalytic validity by demonstrating capability to change participants?
Generalizability	Transferability	Results are useful to other practitioners	Will new concepts be transferable? Is research for practitioners and/or academics?
Objectivity	Permeability/ trustworthiness of researcher	Reflexivity and protection from unwanted bias	Will researcher be open to change provided by the research and will the researcher rigorously test the analysis?

By Rundle (2017) (adapted from Pope et al. 2000)

Wales MBA general manager distributed this survey to a select cohort of residential and commercial builders during a regular member's forum.

The executive director of the third sample cohort from Australia's second peak industry group, the Australian Constructors Association (ACA), advised the author to interface directly with each of its 19 members, through their sustainability sections, to have ACA member's complete questionnaires.

The fourth and final cohort of proponents (comprising of eminent engineering, construction and project management specialists selected by the author) were contacted directly to complete the questionnaires through Qualtrics. As well as the above-noted direct contact between the author and the ACIF, MBA, ACA and the selected eminent peers, on occasion, the author also addressed respondent queries (via telephone and email) regarding the survey questionnaires.

The data for the study was collected using a three-part questionnaire, with Part 1 comprising of 12 close-ended questions relating to sources of waste on project construction sites. A construction waste source identification model developed by Faniran and Caban (1998), based on the seminal work of Spivey (1974), was adopted to ask respondents to indicate their opinion as to the relative significance of construction waste sources. These sources were classified as: (i) design and detailing errors; (ii) client-initiated design changes; (iii) contractor-initiated design changes; (iv) procurement ordering and take-off errors; (v) improper materials handling; (vi) improper materials storage; (vii) poor workmanship; (viii) poor weather; (ix) site accidents; (x) leftover offcuts; (xi) packaging and pallet waste; (xii) criminal waste caused by vandalism or pilfering; and (xiii) lack of on-site materials planning and control. Respondents were asked to indicate the relative significance of the construction waste sources by indicating if the source was "very significant", "significant", "of minor significance" or "not significant" for these 12 questions. In addition, one open-ended question was asked in Part 1, as to other possible "respondent identified" waste sources.

Part 2 of the questionnaire comprised of 10 open questions adapted from a seminal paper by Yates (2013). The first eight questions asked respondents to indicate whether they adopted a particular waste minimization strategy; if this was so, they were then asked to explain, and if not, the respondent was directed to move to the following question. These eight questions concerned: (i) the introduction of training for new techniques to increase efficiency and reduce waste; (ii) the use of innovative designs to reduce waste; (iii) the use of structured approaches for design and construction methods to reduce waste; (iv) as a contractor, do you consciously expend permanent and temporary materials as effectively as possible on behalf of the client; (v) does the contractor use procurement practices as efficiently as possible; (vi) does the contractor use temporary materials as efficiently as possible; (vii) what does the contractor do to minimize material packaging on site (pallets, etc.); and (viii) does the contractor constructively work with the client to minimize scope changes that will possible cause waste.

The respondent was then asked two final questions on Part 2 of the questionnaire, which were: (ix) if their company had a specific site waste minimization strategy in place, and if so, what was it; and (x) the respondent was requested to outline any other possible mitigation strategies.

The third part of the questionnaire entailed the demographic inquiry of each respondent, requesting confidential details, such as their age, gender, professional qualifications, years of employment and size of the largest project they had executed.

5.5.2 Data Analysis

The author used thematic content analysis guided by qualitative description methodology to analyse the Qualtrics survey responses in Part 2 of the survey regarding available strategies to minimize waste. Content analysis was beneficial in the research because it allowed for an unobtrusive method for managing and interpreting descriptive material. This method of analysis enabled the researcher to determine the existence and frequency of concepts in the text. The text was coded, and the common codes formulated the dominant themes. Further sub-themes were developed to explore the dominant themes in more depth. To enhance the credibility of the research, this process was completed by the author and then independently verified by two Ph.D. graduates, who also hold undergraduate engineering science qualifications. Agreement on the main themes was sought between the author and the two specialists through discussion and consensus. Although the qualitative data was analysed in isolation, the overall findings were benchmarked against seminal paper surveys from which these survey questionnaires were prepared.

As noted previously, respondents addressing Part 1 of the questionnaire, which considered possible sources of project construction waste, were asked to indicate the relative significance of the construction waste sources by indicating if the source was "very significant", "significant", "of minor significance" or "not significant". For each construction source, a severity index was determined by calculating the total percentage of respondents giving the response "very significant"; all 12 waste sources were rated. In complete alignment with Faniran and Caban (1998), this subjective respondent assessment was analysed qualitatively, yet using a Likert scale. The Part 1 open-ended question was evaluated using thematic content analysis.

5.5.3 Author Characteristics and Reflexivity Journal

The author has 40 years of experience as a civil engineer, and their highest academic qualification is a Master of Business Administration. He is a project management specialist who has worked long term in 12 different countries. The author has experience in construction execution, design management and engineering studies as chief operating officer, general manager, project director at project manager level down to field level, along with design engineering roles at career commencement. The author has had a career-long interest in executing work more effectively and efficiently and construction waste is one of the four areas where he envisages that long-term benefits can be achieved with minimal effort. The author holds no business or personal affiliations with respondents from any of the three peak body (ACIF, MBA and ACA) member companies undertaking this study. The cohort of engineering, construction and project management specialists who are also undertaking this survey are known to the author on a professional basis only, as past project colleagues from a broad spectrum of construction works. By a process of self-awareness, learned during his Masters of Business Administration studies, the author has endeavoured not to allow any bias to infiltrate the research preparation, data analysis, results and conclusions of this textbook.

A reflexivity journal has been progressively written by the author in accordance with the recommendations by Saldana (2009), to record varying views on coding, themes and sub-themes during the data analysis process. The reflexivity journal documents the author's reflections on the potential research findings to assist in the considerations of emerging themes, concepts and patterns (Creswell and Poth 2017). The reflexivity journal was progressively logged from the coding stage and has also recorded any theme changes. In Chap. 7, "Findings", of this book, only key points from the reflexivity journal are noted against particular survey participant responses. The author also focused on any possible bias, such as confirmation bias, and included these thoughts in the reflexivity register, extracts of which accompany the thematic analysis found in Chap. 7 of this document.

5.5.4 Research Context

The research was carried out for a Masters of Science (Research) Thesis. The author is a civil engineer. The study has been undertaken in Australia with Australian-based companies, but some respondents have related overseas experience.

5.5.5 Sampling Strategy

The sampling strategy is defined as "typical", which provides case information of what is normal or typical (Creswell 2013, p. 158). For phenomenological studies, Polkinghorne (2011) notes that up to 325 proponents can be polled. However, Dukes (1984) argues that, depending on the research, this could be only 10 individuals. The literature review suggests that a sample size for qualitative research should stop at saturation. Authors have suggested that this may occur between 10 (Riemen 1986) and 30 (Creswell 2013) respondents, and this research has adopted these authors' approach. Theoretical saturation of data is a term used in qualitative research and means that researchers reach a point in their analysis of data where the sampling of more data would not lead to further information related to their research questions (Seale 1999).

For the Yates (2013) seminal paper from which this survey questionnaire was adapted to solicit data on site waste minimization strategies, Yates (2013) polled 73 construction executives from the USA's top engineering and construction firms. Twenty-nine of these executives responded with completed questionnaires, which is a 40 per cent response rate. However, as no responses were obtained from the original sample of 150 engineering and construction executives, Yates (2013) altered their strategy by reducing the sample to 73 executives, who were all contacted by telephone prior to receiving the survey questionnaire.

The survey questionnaire from the Faniran and Caban (1998) paper was used to determine the causes of site construction waste for this study and will be used as a benchmark for triangulation purposes to verify data trustworthiness for this research project. Faniran and Caban (1998) sent copies of the questionnaire to 52 construction firms randomly selected from lists of the Master Builders Association of New South Wales and the Australian Federation of Construction Contractors (now defunct). One completed questionnaire per person was returned for the 24 firms who responded, giving a 46 per cent response rate.

For this textbook, the knowledge database was developed on the literature review findings such that the identified front-end construction material wastage process categories would build upon the acquired literature review body of knowledge. Thus, this collated information could be verified or refuted by the survey process. The data acquired in the literature review was critical for assessing the current worldwide contractor operating practices and academic theory on the researched issue. One hundred and two surveys were transmitted to ACIF, ACA, MBA and the selected cohort of engineering, construction and eminent project management specialists. This sample size of nominally 102 survey members complies with saturation (even with the upper limit saturation parameter of 30 respondents) and was expected to supply significantly more completed surveys than both the Yates (2013) and the Faniran and Caban (1998) studies, from which the pro forma questionnaire surveys for this research were adopted.

Coakes and Ong (2011) mention a sample size of 100 or more as being appropriate, albeit for a human factors approach to waste management strategy review, as per the Udawatta et al. (2015) waste management study. Should saturation not be achieved using a sample pool of 100 members, then sampling could readily continue until saturation is attained. For this research, sample participants were asked to provide the completed surveys within two weeks of receiving the questionnaire, but an extra week was allowed, and participants were contacted during this final week, to obtain as large a response as possible.

Tables 5.3 and 5.4 provide details on contractor members from the Australian Constructors Association of our top 19 contracting houses who were polled as sample members, as well as the 12 members from a diverse spectrum of engineering construction companies from the Australian Construction Industry Forum, the peak industry body in this country. As well as these two peak bodies, the Master Builders Association were polled to an undisclosed group of its New South Wales members.

The author endeavoured to mitigate sampling bias by adopting a stratified sampling approach by dividing members of the sample population who have previous

Table 5.3 ACA contractor members

Australian Constructors Association (ACA) Members Group	
No.	Member entity
1	Acciona
2	BGC Contracting
3	Bouyges Contractors Australia
4	Clough
5	CPB Contractors
6	Downer
7	Fulton Hogan
8	Ferrovial Agroman
9	Georgiou
10	Grocon
11	Hansen Yuncken
12	John Holland
13	Lang O'Rourke
14	Lend Lease
15	McConnell Dowell
16	Multiplex
17	Probuild
18	UGL
19	WatPac

By Rundle (2017) (adapted from ACA 2017)

experience in the engineering construction industry, formed into homogeneous sub-groups (Kish 1995), which contain such elements as:

  I. A cadre of top-level management at executive director level who are across their organizations, internationally, as a core respondent cohort, in accordance with Yates' (2013) seminal paper, adopted for Part 3 of the survey questionnaire;
 II. A broad spectrum of management levels drawn from potential respondents, ranging from board chairperson to field superintendent and all of the categories in between (e.g. project manager, site manager);
III. Potential respondents working in the private sector, public sector, academia and as independent consultants;
 IV. Contractors, builders, architects and design consultants;
  V. Clients who execute large capital works programmes;
 VI. Potential respondents who work in heavy infrastructure contracting; process plant installation; mining; commercial building and residential building;
VII. Workers in remote fly in/fly out, inner capital city urban development and large provincial centres, in Australia;

Table 5.4 ACIF Partners Group

ACIF Partners Group		
No.	Partnership entity	Members
1	Air Conditioning and Mechanical Contractors' Association of Australia	Member
2	Australian Consulting Architects	Member
3	Australian Institute of Architects	Member
4	Australian Institute of Building	Member
5	Australian Institute of Building Services	Member
6	Australian Institute of Building Surveyors	Member
7	Fire Protection Association	Member
8	Housing Industry Association	Member
9	Insulated Panel Council of Australia	Member
10	National Fire Industry Association	Member
11	Prefab Australia	Member
12	Society of Construction Law	Member

By Rundle (2017) (adapted from ACIF 2017)

  VIII. Personnel working for Australian companies overseas;

    IX. Landfill operators in brownfield and greenfield;

     X. City engineers in councils with landfill and recycling facilities;

    XI. Project management specialists with experience in the field, in such disciplines as estimating; cost control; planning and scheduling; contracts' procurement and warehousing procurement.

Pursuant to the above, the author's broad stratification of the sample pool has attempted to capture the main areas of work currently being undertaken in the Australian engineering construction industry, comprising: (i) commercial building; (ii) heavy urban infrastructure, such as rail and freeway development; (iii) residential construction; and (iv) mining "de-bottlenecking" brownfield minor capital projects. The recent takeover of Australian engineering construction companies (such as Holland by the Chinese, Leighton by Spanish interests, and SMEC by a Singapore Government consultant) has meant that Australian-based companies have an increased international footprint as a diversification strategy, and that surveys were also provided to Australian companies executing foreign projects (Thompson et al. 2014).

5.5.6 Ethical Issues

Written ethical approval was granted in September 2017 from Southern Cross University for the study questionnaire and accompanying documents, including such issues as respondent confidentiality and questionnaire storage, as a fully compliant study package.

References

Alarcon, L. F. (1994). Tools for the identification and reduction of waste in construction projects. In L. Alarcon (Ed.), *Lean construction*. The Netherlands: A.A., Balkema Publishers.

Alarcon, L. F. (1997). *Lean construction processes*. Chile: Catholic University Press.

Antill, J. M., & Ryan, P. W. S. (1979). *Civil engineering construction*. Australia: McGraw Hill.

Australian Construction Industry Forum (ACIF). (2017). *Latest Forecast*, viewed March to April 2017. https://www.acif.com.au/forecasts/summary.

Australian Constructors Association (ACA). (2017). *Australian Constructors Association Construction Outlook at June 2017*, Australian Constructors Association, viewed June to September 2017. http://www.constructors.com.au/wp-content/uploads/2017/08/Construction-Outlook-June-2017.pdf.

Bloland, H. G. (1995). Postmodernism and higher education. *Journal of Higher Education, 66,* 521–559.

Bossink, B. A. G., & Brouwers, H. J. H. (1996). Construction waste—Quantification and source evaluation. *Journal of Construction Engineering and Management, 122*(1), 55–60.

Cameron, R. (2015). *Mixed methods research workshop, Australia New Zealand academy of management conference*, July 2. Paper presented to Curtin University School of Business, South Australia.

Cherryholmes, C. H. (1992). Notes on pragmatism and scientific realism. *Educational Researcher, 21*(6), 13–17.

Coakes, S. J., & Ong, C. (2011). *SPSS: Analysis without anguish version 18.0 for windows*. Milton, QLD: Wiley.

Conseil International du Bâtiment. (2017). *Home Page*, viewed April to August 2017. http://www.cibworld.nl/site/home/index.html.

Creswell, J. W. (Ed.). (2013). *Qualitative inquiry and research design—Choosing among five approaches*. USA: Sage.

Creswell, J. W., Plano Clark, V. L., Guttman, M., & Hanson, W. (2003). Advanced mixed methods research designs. In A. Tashakkori & C. Teddlie (Eds.), *Handbook of mixed methods in social & behavioral research* (pp. 209–240). Thousand Oaks, USA: Sage.

Creswell, J. W., & Poth, C. N. (Eds.). (2017). *Qualitative inquiry and research design—Choosing among five approaches*. USA: Sage.

Dukes, S. (1984). Phenomenological methodology in the human sciences. *Journal of Religion and Health, 23*(3), 197–203.

Ezo, H. R., Halog, A., & Rigamonti, L. (2017). Strategies for minimizing construction and demolition wastes in Malaysia. *Conservation & Recycling, 120,* 219–229.

Faniran, O. O., & Caban, G. (1998). Minimizing waste on project construction sites. *Engineering Construction and Architectural Management, 17*(1), 57–72.

Ferguson, J. (1995). *Managing and minimizing construction waste: A practical guide*. London, UK: Institute of Civil Engineers.

Formoso, C. T., Isatto, E. L., & Hirota, E. (1999, July). *Method for waste control in the building industry*. Paper presented to the 7th Annual Conference of the International Group for Lean Construction, University of California, Berkley.

Green, A., & May, S. (2003). Re-engineering construction—Going against the grain. *Building Research and Information Journal, 31*(2), 97–106.

Gudmundsson, S. V., & Lechner, C. (2013). Cognitive biases, organization and entrepreneurial firm survival. *European Management Journal, 31*(3), 278–294.

Hao, J. L., Hill, M. J., & Shen, L. Y. (2008). Managing construction waste on-site through system dynamics modelling: The case of Hong Kong. Engineering, *Construction and Architectural Management, 15*(2), 103–113.

Hiete, M., Stengel, J., Ludwig, J., & Schultmann, F. (2011). Matching construction and demolition waste supply to recycling demand: A regional management chain model. *Building Research &* 〓〓〓〓〓〓〓〓 〓〓(〓)〓 〓〓〓〓〓〓

Ibrahim, M. (2016). Estimating the sustainability returns of recycling construction 〓〓〓〓〓 〓〓〓〓〓 building projects. *Sustainability Cities & Society, 23*, 78–93.

Kibert, M. (2008). *Sustainable construction: Green building design and delivery*. USA: Wiley.

Kish, L. (1995). *Survey sampling*. USA: Wiley.

Koskela, L. (2000). *An exploration towards a production theory and its application to construction*. Finland: VTT Technical Research Centre of Finland.

Lincoln, Y., & Guba, E. (1985). *Naturalistic inquiry*. USA: Sage.

Lu, W., Webster, C., Chen, K., Zhang, X., & Chen, X. (2017). Computational building information modelling for construction waste management: Moving from rhetoric to reality. *Renewable and Sustainable Energy Reviews, 68*(1), 587–595.

Maxcy, S. (2003). Pragmatic threads in mixed methods research in the social sciences: The search for multiple modes of inquiry and the end of the philosophy of formalism. In A. Tashakorri & C. Teddlie (Eds.), *Handbook of mixed methods in social & behavioral research* (pp. 51–90). Thousand Oaks, USA: Sage.

Mertens, D. M. (2009). *Transformative research and evaluation*. USA: Guilford Press.

Morse, J. M. (1994). Designing funded qualitative research. In N. Denzin & Y. S. Lincoln (Eds.), *Handbook of qualitative research* (pp. 220–235). Thousand Oaks, USA: Sage.

Nagapan, S., Rahman, I., & Asmi, A. (2012). Factors contributing to physical and non-physical waste generation in construction. *International Journal of Advances in Applied Sciences, 1*(1), 1–10.

Nazech, E. M. (2008). Identification of construction waste in road and highway projects. In: *Paper Presented to the East Asia—Pacific Conference on Engineering and Construction, 2008*.

O'Brien, B., Harris, I., Beckman, T., Reed, D., & Cook, D. (2014). Standards for reporting qualitative research—A synthesis of recommendations. *Academic Medicine, 89*(9), 1145–1151.

Oakley, A. (1999). Paradigm wars: Some thoughts on a personal and public trajectory. *International Journal of Social Research Methodology, 2*(3), 247–254.

Patton, M. (Ed.). (1990). *Qualitative evaluation and research methods*. UK: Sage.

Polkinghorne, D. (2011). Phenomenological research methods. In R. Vale & R. Halling (Eds.), *Existential-phenomenological perspectives in psychology* (pp. 41–46). New York, USA: Plenum Press.

Pope, C., Ziebland, S., & Mays, N. (2000). Analysing qualitative data. *British Medical Journal, 320*, 114–116.

Poulikakos, L., Papadaskalopoulou, C., Hofk, B., Gchhosser, F., Cannone-Falchello, A., Bueno, M. … Partl, M. (2017). Harvesting the unexpected potential of European waste products for road construction. *Resources, Conservation & Recycling, 116*, 32–44.

Riemen, D. (1986). The essential structure of a caring interaction: Doing phenomenology. In P. Munhall & C. Oiler (Eds.), *Nursing research: A qualitative perspective* (pp. 85–105). USA: Appleton-Century-Crofts.

Ritchie, M. (2016). State of waste 2016—Current and future Australian trends. *MRA Consulting Newsletter,* 20 April, viewed March to April 2018. https://blog.mraconsulting.com.au/2016/04/20/state-of-waste-2016-current-and-future-australian-trends/.

Rodrigues, F., Carvalho, R., Evangelista, M., & de Brito, J. (2013). Physical–chemical and mineralogical characterization of fine aggregates from construction and demolition waste recycling plants. *Journal of Cleaner Production, 52*, 438–445.

Rorty, A. (1966). *Pragmatic philosophy—An anthology*. USA: Anchor Books.

Rossman, G., & Wilson, B. (1985). Numbers and words: Combining qualitative and quantitative methods in a single large scale evaluation. *Evaluation Review, 9*(5), 627–643.

Saldana, J. (2009). *The coding manual for qualitative researchers*. Thousand Oaks, USA: Sage.

Seale, C. (1999). *The quality of qualitative research*. UK: Sage.

Song, Y., Wang, Y., Liu, F., & Zhang, Y. (2017). Development of hybrid model to predict construction & demolition waste: China as a case study. *Waste Management, 59,* 350–361.

Spivey, D. (1974). Construction solid waste. *Journal of the American Society of Civil Engineers, Construction Division, 100,* 501–506.

Sugiharto, A., Hampson, K., & Sherid, M. (2002). Non value adding activities in Australian construction projects. In: *Paper presented to the International Conference for Advancement in Design, Construction, Construction Management and Maintenance of Building Products*, Griffith University, Australia, 2002.

Swinburne University. (2016). *How to write a research question*, SU School of Electrical Engineering, viewed March to October 2017. https://www.youtube.com/watch?v=lJS03FZj4K.

Tam, V. (2009). Comparing the implementation of concrete recycling in the Australian and Japanese construction industries. *Journal of Cleaner Resources, 17*(7), 688–702.

Tam, V., & Tam, M. (2006). A review of viable technology for construction waste recycling. *Resources, Conservation and Recycling, 47*(3), 209–221.

Tashakkori, A., & Teddlie, C. (2003). *Handbook of mixed methods in social and behavioral research*. Thousand Oaks, USA: Sage.

Thompson, A., Peteraf, M., Gamble, E., & Strickland, A. J., III. (2014). *Crafting & executing strategy*. New York: McGraw Hill/Irwin.

Treloar, G., Gupta, H., Love, P., & Nguyen, B. (2003). An analysis of factors influencing waste minimisation and use of recycled materials for the construction of residential buildings. *Management of Environmental Quality: An International Journal, 14*(1), 134–145.

Udawatta, A., Zuo, J., Chiveralls, K., & Zillante, G. (2015). Improving waste management in construction projects: An Australian study. *Resources, Conservation and Recycling, 101,* 73–83.

Wahi, N., Joseph, C., Tawie, R., & Ikau, R. (2016). Critical review on construction waste control practices: Legislative and waste management perspective. *Procedia—Social and Behavioural Sciences, 224,* 276–283.

Waste Management World. (2016). *State of the Nation*, Waste Management World periodical, viewed November 2017 to February 2018. https://www.waste-management-world.com/au/report-state-of-waste-2016-current-and-future-trends.

Wu, Z., Yu, A., & Shen, L. (2017). Investigating the determinants of contractor's construction and demolition waste in Mainland China. *Waste Management, 60,* 290–300.

Yates, J. (2008). *Sustainable design and construction for industrial construction*. Austin, TX, USA: The Construction Industry Institute.

Yates, J. (2013). Sustainable methods for waste minimisation in construction. *Construction Innovation, 13*(3), 281–301.

Chapter 6
Results

6.1 Actual Sample Size

One hundred and two surveys were issued to sample members, and 54 respondents replied from 26 October 31 to 30 November 2017, a period of five weeks. Two surveys were undertaken as "test" surveys by two identified respondents, between 22 October 2017 and 24 October 2017, to commission the Qualtrics Web-based system. Over five calendar days from the 26 October 2017, 102 pro forma surveys were progressively transmitted by the author to sample members, with a formal request to undertake the questionnaire. Respondents were formally asked to complete the questionnaire within two weeks from receipt of the survey, but this was extended, so that sample members had five weeks to complete surveys.

Fifty-four respondents completed the Qualtrics survey consent form at the beginning of the questionnaire, but 53 respondents completed survey Part 1 covering participant demographics; 51 respondents completed Part 2 of the survey, which considered sources of site waste; and 48 respondents completed Part 3, which addressed front-end waste reduction strategies. As there was one respondent who only completed the consent form and provided no further input, this book generally quotes the participant response sample as 53, not 54.

A sample submission size of 105 was calculated for the SCU ethics committee form submitted in September 2017 as a deliverable for the 3 October 2017 confirmation approval process, with 102 surveys actually transmitted to possible participants. Guest et al. (2006, p. 59) highlight that the idea of saturation at concept level (from which the original sample size of 102 for this research was calculated back from) "provides little practical guidance for estimating sample sizes for robust research, prior to (actual) data collection". During the 7 November 2017 meeting of the Chair of the SCU Human Research Ethics Committee with the authors sample size and saturation size (30 upper limits, as per Creswell 2013) were confirmed in writing.

The ultimate respondent sample size of 53 participants is a function of the practical constraints of the study duration, allowing one month to receive survey data,

© Springer Nature Switzerland AG 2019   105
P. G. Rundle et al., *Effective Front-End Strategies to Reduce Waste on Construction Projects*, https://doi.org/10.1007/978-3-030-12399-4_6

along with the constructive advice and positive encouragement from the SCU Ethics Committee Chair, in harvesting all the available data from respondents. Accordingly, if 102 pro forma reports were distributed to potential participants in the allotted time-frame, it would serve no benefit to the research to take no account of all respondent data at an arbitrary cut-off point of 30 completed surveys. Thomson (2004) under-took a survey of 50 published ProQuest research papers and found that 22 per cent subscribed to the Morse (1994) saturation criterion of between 30 and 50 responses, whilst 34 per cent of these papers prescribed to the Creswell (1998) criterion of between 20 and 30 for a sample size.

6.2   Completed Survey Questionnaires

Preliminary polling via cold calling of the 19 leading Australian contractor mem-bers of the Australian Constructors Association (ACA) in early October 2017, as suggested by the ACA managing director, indicated that this approach would be unsuccessful. Accordingly, the author prepared a matrix of all colleagues who either worked for, or worked on, projects currently under execution by these 19 top contrac-tors, with surveys sent to 14 of these contractors. The Master Builders Association and Australian Construction Industry Forum issued surveys to selected members as agreed. One of Melbourne's most prolific high-rise commercial builders provided survey input, as did New South Wales and Queensland residential and commercial builders. Western Australian and Queensland mining capital works process Brown-field specialist provided surveys.

The above approach netted 53 survey responses, which was translated into a strong response rate of 52 per cent for the 102 distributed pro forma surveys. Part 1 of the survey questionnaire had a 100 per cent completion rate on most of the questions from the 53 respondents, with slightly lower rates on the perceived possibly "sensitive" questions pertaining to "age" and "years of experience". Part 2, on "Sources of Waste", had 51 responses that were 100 per cent completed. Part 3, on "Possible Strategies to Reduce Waste", had 48 responses that were 100 per cent completed.

As can be seen from the following demographics summation of respondents, 16 industry company directors (including a chair, two CEOs and four Project Direc-tors) completed the survey, as well as two Professors of Civil Engineering with field experience; this explains why 30 respondents had 26 years or more experience and 24 respondents held postgraduate degree qualifications (masters or Ph.D.). Notwith-standing this top management decision-making cadre, there were eight respondents with trade vocational qualifications and a building licence who provided expert advice from a superintendent and foreperson perspective. Forty-six males and seven females completed the survey. A range of "other" construction engineering specialists also completed the survey (project managers, site managers, construction managers, risk managers, sustainability managers, estimators, schedulers, business developers, engineering design managers, architects, planners, landfill managers, city engineers, environmental managers, quality managers, contract managers and clients).

The 52 per cent response rate was a good result in the light of the response rate to other papers on waste management, such as Yates (2013), upon which Part 3 of this survey questionnaire was based, who originally polled 150 construction and engineering executives from the peak industry body funding their research; zero responses were received from these Construction Industry Institute members. Subsequently, repeat polling commencing with a telephone discussion of 73 engineering and construction top-level executives resulted in 29 completed surveys for a 40 per cent response rate, in line with the Creswell saturation general upper limit of 30 (Creswell 1998; Creswell and Poth 2017).

The Faniran and Caban (1998) study, for which Part 2 of this questionnaire survey was adopted, used a sample size of 52 companies from the Master Builders Association and the Australian Construction Federation and received 24 responses, representing a 46 per cent participant rate.

6.3 Survey Saturation

The 102 surveys provided to prospective sample members were predicated on obtaining at least 25 respondents at a predicted 25 per cent response rate, which would comply with Charmaz's (2006, p. 114) saturation level of 25 respondents and (in view of the phenomenology philosophical approach underpinning the pragmatic framework) a maximum of 25 respondents as a specific phenomenological approach by Creswell (1998, p. 64). As noted in Sects. 6.1 and 6.2, an empirically derived upper limit saturation of between 25 and 30 respondents was targeted for this research with 53 responses received. Even though Part 1 of the demographic survey questions had some age sensitive data missing, Part 2 had 51 complete responses and Part 3 provided 48 complete responses. All of the written responses from the entire 53 received responses were harvested. Only the subjective quantitative data in Chap. 2 was calculated using a base figure of 51 respondents. The harvesting of all data, including respondent remarks on partially completed surveys, was broadly discussed with the Chair of the SCU Human Ethics Committee and was discussed and confirmed in writing on 7 November 2017.

Mason (2010, Note No. 14) posits that:

> There is an obvious tension between those who adhere to qualitative research principles, by not quantifying their samples—and those who feel that providing guidance on sample sizes is useful. Some researchers have gone further than providing guidelines and have tried to operationalise the concept of saturation, based on their own empirical analysis.

Dey (1999) argues strongly that the concept of saturation is inappropriate because researchers too often close categories prematurely, using partially coded data. Strauss and Corbin (1998) suggest that saturation is a matter of degree, where often researchers demand too much data to allow potential new outlier information to emerge, whilst the primary concern of data collation should be that sufficient information needs to be collected so that it is counterproductive to continue to evolve the overall story, model, theory or framework. Strauss and Corbin (1998) believe too much data can even confuse the researcher and make it difficult to derive meaningful research conclusions.

Morse (1995, p. 147), in an editorial for the *Journal of Qualitative Health Research,* robustly defends the position that the researcher is obligated to leave no stone unturned, especially during the early study phase, to uncover all possible data related to the research topic, including any outlying information which may be perceived as "insignificant". This is because the richness of data is the ultimate aim of attaining saturation, contrary to attaining an empirically derived number, or contrary to the common saturation concept of undertaking an analysis of the data, until "frequency of occurrence has occurred" (Morse 1995, p. 147).

Pursuant to the above, it can be seen why Mason (2010) and Strauss and Corbin (1998) suggest that two of the most common postgraduate research questions posed to supervisors. They are: (i) How to determine that there is sufficient data, and (ii) how large should the survey sample be?

Mason (2010) has researched the issue of qualitative research sample sizes/saturation for Ph.D. students. A review of postgraduate research entry requirements for the globe's top 50 universities indicates that funding sponsors generally impose both time and financial constraints on student applications that specifically limit research sample sizes. As an example, the Toronto University Ph.D. programme asks the candidate to "justify anticipated sample size", whilst the potential Ph.D. candidate at the University College of Dublin is asked to "give an example of sample size and selection" (Mason 2010, p. 5). Part 2 of the SCU Fast Track Ethics Committee approval pro forma asks for the postgraduate research candidate to provide the "intended number of survey participants".

Mason (2010, Table 1.1) executed a content analysis of 532,646 higher degree research project abstracts in Great Britain and Ireland from 1716 to 2009 and found that for all phenomenological qualitative research projects, the average participant sample size was 25.

Pursuant to the above, the author has adopted an empirical approach to determining saturation. Section 6.3 has comprehensively shown that a sample size of 53 respondents with a minimum of 48 completed surveys exceeds the minimum requirements. Nevertheless, the author after duly interpreting the data via coding, and developing themes and sub-themes, shall further comment in Chap. 7, "findings", on the theoretical saturation concept approach to classic qualitative research. This is as per the instructions of Morse (1995, p. 147) who defines saturation as having "'data adequacy' and operationalized as collecting data until no new information is obtained", which is in alignment with the seminal saturation approach to sampling outlined by Glaser and Strauss (1967).

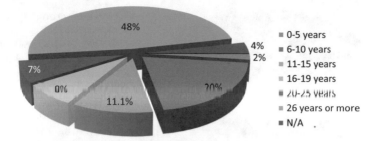

Chart 6.1 Respondent engineering construction work experience

6.4 Respondent Demographics

There were 53 respondents captured in these demographic results. When there was an unanswered respondent question, it has been noted in the following tables as N/A (not applicable).

6.4.1 Respondent Gender

Forty-six males and seven females completed surveys; this figure, 13.2 per cent of the respondents, approximates the percentage of female professionals currently engaged in the Australian engineering construction industry of around 14 per cent (UNSW 2016).

6.4.2 Respondent Engineering Construction Work Experience

See Chart 6.1.

6.4.3 Value of Largest Project Respondent Worked On

See Chart 6.2.

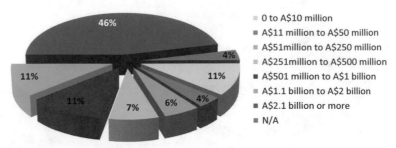

Chart 6.2   Value of largest project respondent worked on

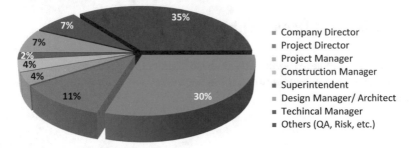

Chart 6.3   Current position of respondent

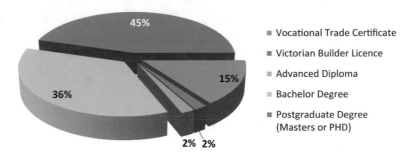

Chart 6.4   Highest academic qualification held by respondent

6.4.4   Current Position of Respondent

See Chart 6.3.

6.4.5   Highest Academic Qualification Held by Respondent

See Chart 6.4.

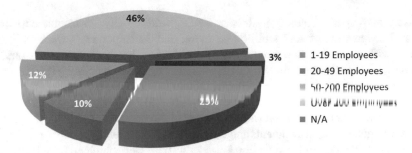

Chart 6.5 Number of employees in respondent company

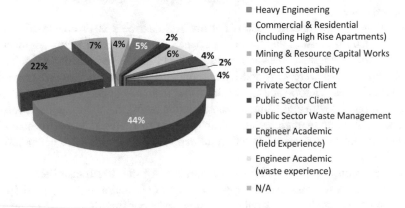

Chart 6.6 Respondent field of engineering construction work

6.4.6   Number of Employees in Respondent Company

See Chart 6.5.

6.4.7   Respondent Field of Engineering Construction Work

See Chart 6.6.

6.4.8   Summation of Respondent Demographics

The high calibre of the top managerial cohort held by 16 respondents at board and executive level, along with six project directors, out of the 53 survey participants, explains the large number of postgraduate qualifications attained by the respondents, with 24 participants holding masters or Ph.Ds. Additionally, 25 respondents have worked on projects of A\$2.1 billion or more, and 32 respondents work for companies with more than 200 employees. Valuable data has been gathered from the above-noted top-level engineering construction executives who are very aware of their organization's strategic imperatives, including profit and the impact that corporate social responsibility issues regarding sustainability and ecological matters have on their firm's reputation. Data was also sourced directly from the field, via project managers, general superintendents and field supervisors. In addition, advice and opinions to add to the research acquisition knowledge base were obtained from local government waste management officers; public and private sector clients; project management specialists; architects/design managers; and engineering academics with field experience.

A review of the Australian Constructors Association's 2017 construction outlook for Australia showed that the split between respondents for heavy engineering construction, mining resources and building disciplines is reasonable (ACA 2017). Completed surveys were provided by respondents working in Western Australia, New South Wales, Victoria and Queensland and from respondents working for Australian companies, with international reach ranging from USA to Mongolia. Overseas business by Australian engineering construction firms represented an estimated 2.7 per cent of total Australian construction industry turnover in the 2017 calendar year (ACA 2017).

In conclusion, the 53 respondent samples, out of the 102 transmitted surveys, represents a diverse cross section of engineering construction industry participants that should provide a broad, yet balanced view of the construction site waste matters considered in this research survey.

6.5 Sources of Construction Site Waste Survey Results

Section 6.5 considers research results of respondent survey answers to Likert scale Questions 12–24 inclusive, which evaluated 13 potential site waste sources on Australian construction projects based on the Faniran and Caban (1998) survey questionnaire used for this research. The respondents were asked to evaluate the 13 potential sources of waste as being either: (i) "not significant"; (ii) "of minor significance"; (iii) "significant"; or (iv) "very significant". Acknowledgement is made of Faniran and Caban's (1998) work, developing the earlier seminal study of Spivey (1974) and Gavilan and Bernold (1994) into a construction waste source identification model, alluded to in many subsequent papers including the Nagapan et al. (2012) paper (which has used an abridged model); their research was used for benchmarking purposes with this research. The Faniran and Caban (1998) approach to ranking the order of importance of the 13 potential sources of waste on the survey questionnaire, via severity index, has also been adopted in this research.

This research adopted two refinements to Faniran and Caban's (1998) construction waste source identification model, in the research questionnaire Part 2, "Sources of Site Waste", as follows:

I. The "design changes" source in the Faniran and Caban (1998) model was split into two questions, comprising "client-initiated design changes" and "contractor-initiated design changes". The reason for this amendment was that recent literature indicated that client-initiated changes are a significant source of site waste because they create redundant materials due to a scope change (Yates 2013). Accordingly, it was important to categorize this source option into two potential sources, which shall be seen in the data appraisal. This resulted in two very different views on design change source causation and has had a major bearing on these research findings, discussion and conclusions.

II. A further question was asked to the respondents to solicit their opinions, based on their field of engineering construction experience, on any other possible sources of waste further to the 13 options denoted in Part 2 of the questionnaire. A thematic analysis was selected to evaluate the rich harvest of data obtained from this valuable question.

Part 2 of the research questionnaire evaluated 13 potential sources of construction site waste which were classified as: (i) design and detailing errors; (ii) client-initiated design changes; (iii) contractor-initiated design changes; (iv) procurement ordering and take-off errors; (v) improper materials handling; (vi) improper materials storage; (vii) poor workmanship; (viii) poor weather; (ix) site accidents; (x) leftover offcuts; (xi) packaging and pallet waste; (xii) criminal waste caused by vandalism or pilfering; and (xiii) lack of on-site materials control and planning. Fifty-one respondents correctly completed Part 2 on sources of construction waste, and therefore, this was the respondent sample size adopted for this research section, as per the data provided by the Qualtrics software.

6.5.1 Severity Index

Respondents were asked to indicate the relative significance of construction site waste sources by specifying if the sources were "very significant", "significant", "of minor significance" or "not significant". In accordance with the Faniran and Caban (1998) model, (from which Part 2 questionnaire for this research was adopted) for each of the 13 construction site waste sources identified on the survey, a severity index was determined by calculating the percentage of respondents giving the response "very significant" and the 13 site waste sources were ranked on this basis. Table 6.1 shows the severity index and the ranking for each source of construction site waste.

Table 6.1 Potential sources of construction waste severity index

Survey question number	Possible source of site waste	Severity index = # "very significant"/total respondents (%)	Severity index ranking
24	Lack of on-site material planning and control	24	1
22	Packaging and pallet waste	20	2
12	Design and detailing errors	20	2
13	Client-initiated design changes	20	2
17	Procurement ordering and take-off errors	16	5
16	Improper materials storage	14	6
18	Poor workmanship	14	6
15	Improper material handling	10	8
14	Contractor-initiated design changes	8	9
20	Site accidents	4	10
21	Leftover offcuts	4	10
19	Poor weather	2	12
23	Criminal waste caused by vandalism or pilfering	2	12

By Rundle, November 2017 (adapted from Faniran and Caban 1998)

6.5.2 Ranking 13 Sources of Construction Site Waste

The author has also opted to use a second alternative method, in order to evaluate a ranking for the potential 13 sources of construction site waste in Questions 12–24, inclusive, of Part 2 of this research questionnaire. This is due to the close scoring by using the severity index approach to rank the 12 source options, to determine if an alternative methodology in synthesizing data shall provide a clear ranking of possible sources. Accordingly, scores were given for each response to all 13 questions as follows: (i) "very significant" was scored a three; (ii) a "significant" answer was given a two; (iii) a response given as "of minor significance" was rated as a one; and (iv) a "not significant" reply was given a zero score. Table 6.2 provides ranking

Table 6.2 Ranking of potential sources of construction waste by scoring each response

Survey question number	Possible source of site waste	Respondent scores 0 = not significant 1 = limited significance 2 = significant 3 = very significant	Respondent score/maximum score (maximum possible score = 153) (%)	Ranking of sources using all responses
12	Design and detailing errors	110	72	1
13	Client-initiated design changes	91	60	2
22	Packaging and pallet waste	90	59	3
24	Lack of on-site material planning and control	88	58	4
15	Improper material handling	83	54	5
17	Procurement ordering and take-off errors	82	54	5
16	Improper material storage	81	53	7
18	Poor workmanship	79	51	8
21	Leftover offcuts	75	49	9
14	Contractor-initiated design changes	58	38	10
20	Site accidents	49	32	11
19	Poor weather	47	31	12
23	Criminal waste caused by vandalism or pilfering	25	16	13

By P. Rundle, December 2017

of these 13 sources of construction waste scoring each Question 12–24, inclusive, pursuant to the above.

When comparing Tables 6.1 and 6.2, the severity index ranking method and the scoring all responses method produced the same five of the six top major sources of waste, but not in the same order. Further, for the four sources deemed least likely to cause construction waste, three of these bottom source selections were the same, for Tables 6.1 and 6.2.

6.5.3 Respondent Comments Covering Other Sources of Waste

Question 25—Part 2: Sources of Waste
Part 2 of this research survey questionnaire asks respondents in Questions 12–24, inclusive, to address a Likert scale evaluation on the 13 most commonly recognized sources of construction waste on project sites, based on a review of academic literature, including seminal papers. The final question, Question 25 of Part 2, requests that respondents: *"list any other sources in your opinion that cause site construction waste"*. In strict accordance with the Faniran and Caban (1998) model, though a Likert scale approach was adopted to obtain data, respondent's opinions were sought. The research methodology was qualitative; therefore, no statistical analysis was undertaken. Table 6.3 captures all respondent comments and opinions regarding other possible sources of construction waste on Australian projects.

6.6 Waste Minimization Strategies Survey Results

This section provides results from the survey questionnaire Part 3, Questions 26–34, inclusive, in which respondents were asked to answer questions on the use of waste minimization strategies derived from the Yates (2013) survey questionnaire (duly amended to include those questions that related to front-end site waste minimization strategies). This was to determine from the 48 participant responses whether these initiatives are being adopted on Australian projects. The focus of this questionnaire was on upstream site reduction strategies, rather than downstream construction materials supply chain considerations, such as recycling of waste materials and/or transport of residual waste to landfill. Respondents could answer "yes", "no" or "don't know". If respondents answered in the affirmative to using one of the strategies nominated in Question 26–33, inclusive, then further comments were solicited. Question 34 requested respondents to provide information on other waste minimization strategies adopted on their projects over and above those nominated in Questions 26–33, inclusive.

Acknowledgement is given to Yates (2013) with respect to Part 3 survey questions on waste minimization used in this research. An abridged form of Yates' (2013) research survey was used. None of Yates' (2013) questions on recycling were included in Part 3 of the survey because only front-end waste reduction practices were considered in the author's research. Yates' (2013) approach was to tabulate

Table 6.3   Respondent's opinions on sources of site waste

No.	Respondent's opinions on possible sources of waste, over and above questionnaire Part 2, Questions 12–24, inclusive
1	Inaccurate procurement processes that cause site construction waste
2	Generally good planning, procurement and delivery schedules
3	Packaging and temporary formwork(s) for poor quality
4	Overengineering by designers (lack of optimization and lack of value engineering)
5	Incorrect classification of waste (e.g. classifying something to be sent to landfill when it actually could have been re-used or recycled); lack of understanding about local waste classification and guidelines; not knowing what materials can be recycled/re-used
6	Poor segregation (of construction waste)
7	Vendors
8	Poor housekeeping
9	All major causes covered in above Question 12 to Question 24 complete
10	Poor planning
11	Lack of proper supervision
12	Inability to recycle certain plastics and concrete surplus often results in waste to landfill
13	Logistics of concrete deliveries
14	Poor overall management and its enforcement by key site personnel
15	Client review and approval of management plans and design submissions
16	Inadequate clean-up of work areas on a continuous basis
17	Client requirements for samples and prototypes
18	Insufficient storage facilities, leading to spoiling of materials (e.g. bags of cement left in the open)
19	Site clean-up at end of project
20	Oversights in planning leading to rework and redesign, to secure relevant approvals before start of project or to progress on new stage/extension
21	Work culture not valuing materials
22	No consideration of construction waste minimization as a primary design feature
23	Individual trades not being held accountable for the waste they generate
24	Oversupplied materials
25	Engineers and architects
26	Incorrect estimation of material
27	Defective materials or workmanship
28	Poor personal hygiene practices by sub-contractors/workers and poor clean-up, as project progresses
29	Lack of attention to early design effort
30	Overordering and poor handling
31	Lack of suitable outlets; space limitations on site [e.g. space to segregate and sort; timing of waste production versus availability of recyclers (inert material)]
32	Ten per cent wastage adds up to a significant factor in construction wastage

(continued)

Table 6.3 (continued)

No.	Respondent's opinions on possible sources of waste, over and above questionnaire Part 2, Questions 12–24, inclusive
33	Poor workmanship causing breakout work; lack of concern by section engineers and groups for wastage and primary materials
34	Poor supervision of materials control, at installation/build point
35	Mostly poor quality of materials and/or equipment, from China and other Asian locations
36	Not engaging (client) operations, early in the design
37	Insufficient waste collection and waste (storage) facilities. Uncovered skips
38	Lack of planning and actively promoting recycling and re-use of "waste" materials for "beneficial" use of local communities
39	Engineering designers and architects do not design to mitigate waste
40	Contractors pass on material waste costs to clients. Clients are a source of waste by not configuring a mandatory waste reduction plan in contract, as part of its tender evaluation
41	Not specifying recyclable materials, such as filters (in the permanent works), during the design and maintainability baseline development
42	No segregation for recycling
43	Allowing "overcontingency" when ordering

By P. Rundle, December 17

comments from respondents for analysis. Part 3 questionnaire adapted from Yates (2013) and used in this research was:

> designed to include questions that could be answered with "yes", "no", or "do not know" responses and if the respondents answered in the affirmative they were provided with additional space to elaborate on their "yes" answer by providing examples of situations, where methods, processes, or ideas were implemented on actual construction projects. The questionnaire included check boxes for answers along with textboxes for written clarification of the checkbox answers (Yates 2013, p. 284).

The Yates (2013) qualitative method of data synthesis was to tabulate the Part 3 question responses into the most common/least common site waste minimization processes and to tabulate respondent remarks to any "yes" answers. From this, a summary of the most commonly employed methods of site waste minimization in the USA was obtained. Several of the "senior executive respondents had 30–40 pages of written responses to questions" (Yates 2013, p. 297). This research followed the same data synthesis approach as Yates (2013), up to a point. All data was tabulated with respondent's answers summarized, and their comments to "yes" answers are also tabulated. However, in-depth data synthesis in this textbook was undertaken by thematic analysis to determine the findings.

Thematic analysis focused on evaluating the themes captured within the data (Creswell et al. 2003). Coding is the main process for preparing and developing themes from the raw data (Creswell et al. 2003). Thematic analysis is consistent with the phenomenology approach adopted for this research, which focuses on the human subjective experience that has emphasized participant's perceptions, opinions and experiences on the topic (Creswell and Poth 2017).

6.6.1 Respondent Survey Data on Waste Minimization Strategies

Table 6.4 captures answers from 48 respondents to Questions 26–34, inclusive, covering waste minimization strategies, whether in the affirmative, negative or simply "don't know". This approach follows Yates' (2013) methodology used to record data results from the survey questionnaire, also used by this research duly abridged, as detailed in Sect. 6.6.

6.6.2 Respondent Comments on Waste Minimization Strategies

This section provides details of respondent comments, if answering in the affirmative to Questions 26–33, inclusive, on waste minimization strategies. Comments from Question 34 are also included, which requests respondents to remark on other possible waste minimization strategies, over and above those nominated in Questions 26–33, inclusive.

6.6.2.1 Respondent Comments Following a "Yes" Answer to Question 26

Question 26—Part 3 Waste Minimization: Is your firm using techniques that improve resources efficiency, equipment efficiency and material resource efficiency and allow for the training of manual labour? (Table 6.5).

6.6.2.2 Respondent Comments Following a "Yes" Answer to Question 27

Question 27—Part 3 Waste Minimization: Are innovative designs, construction components or construction processes being integrated into your projects to reduced site generated waste? (Table 6.6).

Table 6.4 Respondent answers to Part 3 site waste minimization strategies, Questions 26–34

Part 3 waste minimiza- tion question no.	Question description	% Respon- dents answered "no"	% Respon- dents answered "don't know"	% Respon- dents answered "yes"
26	Is your firm using techniques that improve resources efficiency, equipment efficiency, material resource efficiency and allow for training of manual labour?	39.58	18.75	41.67
27	Are innovative designs, construction components or construction processes, being integrated into your projects to reduce site generated waste?	31.25	20.83	47.92
28	Do you adopt a structured approach both for engineering design and in determination of construction methodologies that involve waste minimization strategies?	29.17	20.83	50.00
29	Do you address waste generation reduction during project pre-planning to utilize designs that minimize waste using any of the following techniques: precast; prefabrication; pre-assembly and modularization?	29.17	14.58	56.25
30	As builders, contractors and engineering consultants, do you ensure a minimum amount of permanent and temporary materials are expended in the effective provision of client conforming construction/building product	31.25	31.25	37.50
31	Regarding use of temporary construction materials, do you consider waste minimization processes? As an example, for concrete construction, do designers specify concrete elements of similar dimensions, where practical? Are steel shutters used on repetitive formwork; is formwork adequately treated and robustly fabricated to allow re-use; are orders "just in time", to reduce material losses on site?	18.75	29.17	52.08
32	Do the contractor/builder, consultants and vendors constructively work with the client to minimize change orders that make pre-ordered products, redundant and suitable only for waste?	35.42	22.92	41.67
33	Does the contractor/builder and/or client have a mandatory waste minimization plan developed as part of the project execution plan?	33.33	29.17	37.50

(continued)

Table 6.4 (continued)

Part 3 waste minimization question no.	Question description	% Respondents answered "no"	% Respondents answered "don't know"	% Respondents answered "yes"
34	Provide other examples of situations where methods, processes or ideas were implemented on your construction site projects that minimized waste	16.67	33.33	50.00
Total of respondent answers		264.59	220.82	414.59
Average for each respondent answer		29.40% = "No"	24.54% = "Don't Know"	46.07% = "Yes"

By P. Rundle, December 2017

6.6.2.3 Respondent Comments Following a "Yes" Answer to Question 28

Question 28—Part 3 Waste Minimization: Do you adopt a structured approach both for engineering design and in the determination of construction methodologies that involves waste minimization strategies? (Table 6.7).

6.6.2.4 Respondent Comments Following a "Yes" Answer to Question 29

Question 29—Part 3 Waste Minimization: Do you address waste generation reduction during project pre-planning to utilize designs that minimize waste using any of the following techniques: precast; prefabrication; pre-assembly or modularization? (Table 6.8).

6.6.2.5 Respondent Comments Following a "Yes" Answer to Question 30

Question 30—Part 3 Waste Minimization: As builders, contractors and engineering consultants, do you ensure that the minimum amount of permanent and temporary materials are expended in the effective provision of client conforming construction/building products? (Table 6.9).

Table 6.5 Part 3 Question 26 respondent comments

No.	Respondent comments to Question 26—Part 3 waste minimization
1	Machine systems control
2	Planning and scheduling latest techniques utilized, e.g. use barges for spoil haulage to minimize truck movements in the central business district (CBD)
3	We are (our company) more and more improving design detailing from project lessons learned, to make sure rework is minimized and waste from error is reduced
4	Digital building information modelling (BIM) engineering
5	Multiple skips for sources of separation
6	Very early involvement in the design and off-site fabrication
7	Reviewing site collaboration software to enhance communication and timeframes for accessing information
8	Rigorous inspections at all stages; realistic material delivery schedules
9	Apps on mobile devices to replace paperwork for surveillance; assessment; health, safety and environmental (HSE)
10	Using proper estimating system (like D'Est software) provides immediate insight into material requirements, through the procurement sector of programme
11	Better quality trade and supervision reduce mistakes and waste
12	Better supply chain management
13	Off-site modularization
14	The company I work for uses excellent waste strategies in waste reduction, most times. Recycling bins of exact metal types and the consolidation of (scrap) large nuts, bolts, studs, etc.
15	We are designers and site supervisors and do not have a direct effect on construction materials procurement, delivery and utilization/construction, to manage or affect contractor waste
16	Integrated supply chain. "Just-in-time" deliveries
17	Equipment selection; training; process planning and incentives
18	Recommending to clients to include a procurement baseline from concept through to commissioning that includes sustainable use of materials, in both construction and ongoing maintainability

By P. Rundle, January 2018

6.6.2.6   Respondent Comments Following a "Yes" Answer to Question 31

Question 31—Part 3 Waste Minimization: Regarding the use of temporary construction materials, do you consider waste minimization processes? For example, for concrete construction, do designers specify concrete elements of similar dimensions where practical; are steel shutters used on repetitive formwork; is formwork adequately treated and robustly fabricated to allow re-use; and are orders "just in time", to reduce material losses on site? (Table 6.10).

Table 6.6 Part 3 Question 27 respondent comments

No.	Respondent comments to Question 27—Part 3 waste minimization
1	Embedding into the Design Management Plan
2	Promoting single package bundling of multiple components rather than multiple, Individual wrapping component are Pulling and interim storage in containers to provide protection until use—results in reduction of weatherproof packaging
3	Off-site fabrication in factory environment where possible
4	Modular construction strategy
5	Yes, but not directly to reduce waste but to provide an efficient design. It is assumed that the contractor will manage waste as this would be reflected in construction cost and time reductions
6	Limited scale depending on contract (e.g. are alternative design/solutions accepted) and alternative component approval processes
7	Off-site fabrication; quality detailed design; good supervision; planning
8	Lost formwork, integrated structural elements, bond deck
9	As per Question 26 above
10	Being aware of potential waste generation and taking the time to minimise it
12	Modularization and maximised pre-assembly and precast are the key to reduced construction waste
13	Approval documents are being developed to focus on enabling companies to maximise opportunities to use the most efficient designs and construction methodologies by focusing compliance requirements on the outcome to be achieved rather than the process to achieve compliance
14	Typically driven by constructor initiatives' prior design finalization. Significant room for improvement by design consultants
15	Utilization of Davit and tower cranes
16	We are constantly looking for innovative design approaches to give us an advantage and that includes minimizing wasted time
17	Greater use of prefabrication off-site
18	Very early involvement in the design, input in the design company experts, implementation of off-site fabrication and partial off-site pre-assembly of building elements
19	Ground stabilization product that enables us on "unsuitable" engineering soil as useable material
20	As per above, innovative design for us means "smarter and easier" construction in the field and this reduces waste through error in a tough and competitive commercial building sector in Melbourne
21	3D design tools
22	Seeking resource recovery exemptions to maximise the amount of materials that can be reused; recycling targets set in contract requiring high percentage of construction and demolition waste to be recycled and spoil to be reused; negotiate take-back agreements with suppliers for packaging; maximise use of recycled or reused materials (e.g. wood, steel and site facilities)
23	We have a waste minimization policy which is in development

By P. Rundle, January 2018

Table 6.7 Part 3 Question 28 respondent comments

No.	Respondent comments to Question 28—Part 3 waste minimization
1	Design and engineering use the latest information in the industry
2	We have waste management plans, but their scope is limited to contractor activities
3	Early contractor involvement (ECI)
4	Shipping container contents are considered during engineering
5	Attention to detail in design and shop detailing
6	Structurally efficient
7	We pay considerable attention to early design in the definitive stage
8	Value engineering in design
9	For design and build contracts where there is economic incentive to design out waste or in response to sustainability rating schemes (green building/infrastructure)
10	Constructability reviews
12	Part of method statement development
13	I think this emanates from PM, quality management system (QMS), staging and construction planning
14	Is taken into consideration
15	A project-specific planning assessment as to programme preparation and product availability
16	Thoughtful design development and documentation
17	We have a structured approach—it is immature, and we are starting to embrace LEAN technologies to improve our approach
18	Planning of tunnel spoil disposal in landfill and balancing cut and fill in road construction
19	Maximise employment of full pre-assembly
20	Regular design and implementation reviews. Running of a risk and opportunity register
21	Embedded into the Design Management Plan

By P. Rundle, January 2018

6.6.2.7   Respondent Comments Following a "Yes" Answer to Question 32

Question 32—Part 3 Waste Minimization. Do the contractors/builders, consultants and vendors constructively work with the clients to minimize change orders that make pre-ordered products redundant and suitable only for waste? (Table 6.11).

Table 6.8 Part 3 Question 29 respondent comments

No.	Respondent comments to Question 29—Part 3 waste minimization
1	Yes. Precast is used where appropriate and steel table formwork/falsework for high-rise soffits
2	Yes previously fabricate units and fabrication of large electrical and mechanical units/products in factories/manufacturing facilities
3	We definitely are being asked to consider more modularised options for our engineering and architectural designs in the industrial sectors for manufacturing and mining
4	Used intermediate bulk containers are donated to people requiring water storage off-grid
5	Use prefabrication where possible
6	Strong drive for prefabrication, pre-assembly and modularization, e.g. prefabrication high-voltage (HV) sub-stations delivered to site
7	String focus on modularization and off-site pre-assembly
8	Prefabrication and pre-assembly—up to the Subbie to manage waste
9	Prefabrication is maximised to the limits of transport
10	Prefabrication and modularization used, selection is project dependent and varies between projects
12	Precast, prefabrication and modularization
13	Precast/modular construction techniques; site measurement
14	Precast or modularization is commonly used
15	Off-site prefabrication and modularization
16	Off-site construction is more efficient than onsite
17	Most projects push the responsibility of waste management to EPC or EPCM, who in turn push it further down the commercial pecking order, and finally to landfill.
18	Modularization and bulk packaging
19	Identification of suitable spoil disposal areas
20	Current project utilizing precast components, including precast factory onsite
21	Construction methodology well done by experienced people
22	Constantly looking to minimise waste—precast, retain existing infrastructure, avoid temporary works and side—tracks, etc
23	Assessing the design—design to suit logistics (roads transport/lifting capacities), implementation of off-site manufacturing, off-site partial pre-assembly

By P. Rundle, January 2018

6.6.2.8 Respondent Comments Following a "Yes" Answer to Question 33

Question 33—Part 3 Waste Minimization: Does the contractor/builder and/or client have a mandatory waste minimization plan developed as part of the project execution plan? (Table 6.12).

Table 6.9 Part 3 Question 30 respondent comments

No.	Respondent comments to Question 30—Part 3 waste minimization
1	Minimizing materials minimises cost
2	Waste minimization and sustainable materials strategies implemented on project
3	Part of our sustainability model
4	Minimise overorder and retain excess for inclusion in the next job
5	Sometimes specify recycled materials, e.g. glass sand
6	Methodology of erection/temporary works are designed at the same time as permanent elements of the building
7	If we do not do this, then we do not win work
8	Not sure as to what extent
9	Waste minimization is part of conscientious cost control
10	By conducting value engineering exercises
12	Usually, anything temporary is found a new "home/purpose", somewhere else in the project
13	As per Question 28, above
14	Prefabrication and quality installation contracts
15	Actions are taken to minimise material use; often, this is outside of our control and is the building contractor. This often results in designs presented for approval which could be modified to reduce waste, but time constraints prevent the additional cycle to modify the design
16	Where there is opportunity through the contract and mutual benefit (e.g. green building)
17	We deliver to the client a healthy standard of workmanship and minimal wastage to minimize actual costs

By P. Rundle, January 2018

6.6.2.9 Respondent Comments Following a "Yes" Answer to Question 34

Question 34—Part 3 Waste Minimization: Provide other examples of situations where methods, processes or ideas were implemented on your construction site projects that minimized waste (Table 6.13).

Table 6.10 Part 3 Question 31 respondent comments

No.	Respondent comments to Question 31—Part 3 waste minimization
1	To an extent—it is risky to run full "just in time" (JIT)
2	Maximise use of recycled or reused materials (e.g. wood, steel, site facilities)
3	We use the most cost effective accurate production techniques that also consider worker knowledge of method: precast, steel shutters, etc. Quick construction, with formwork maintains profits, finishes projects on time and reduces waste
4	Reuse formwork where possible
5	Have previously developed reusable forms for spread footings and formwork
6	By use of available formwork systems
7	Very tight and accurate programming
8	Always looking to simplify and avoid temporary works costs that do not add value
9	This is standard practice
10	By specifying the type of formwork to be used
12	Standardization is virtually non-existent. Precast is almost an evil conjuring. How could the large engineering houses make profit from cookie cutter approach. The baulk at pre-assembly because it minimises post-design-phase follow-up works
13	Dictated by the main contractor (MC) or contractor generally
14	Prefabrication wherever possible. Delivery "just in time". Construction planning
15	As a Project Management Consultant (PMC), these are not directly controlled or specified in contracts. However, selected sub-contractors (e.g. concrete construction) do make use of reusable formwork
16	Standardization of design elements. Reuse of temporary construction materials
17	Where feasible, often clients design may not suit that approach (non-standard elements, etc.)
18	Yes and no, depending on the job contractor and or client in certain situations, yes in others quite the opposite
19	Yes—repetitiveness but for cost efficiency resulting in re-use of temporary works
20	"Just-in-time" orders, standardization of concrete structures
21	Permanent formwork, steel and aluminium scaffold re-use
22	Reuse of formwork and falsework critical to prudent temporary works design and installation for commercial reasons

By P. Rundle, January 18

Table 6.11 Part 3 Question 32 respondent comments

No.	Respondent comments to Question 32—Part 3 waste minimization
1	Many coordination meetings and training
2	N/A to current project
3	Through collaboration "from day one"
4	Where possible, we will aim to have weekly interaction meetings to keep things on track and focused
5	"Managing the client" is still a challenge
6	If obliged to do so in accordance with the consultants' technical specification
7	Any variation has to be carried out at lowest cost. The contractor does not overorder, as the client usually will not pay reasonable costs in every case
8	Through more efficient contractors
9	This is facilitated by establishing good, interactive, regular communication with the client, where programme and staging are discussed
10	Regular meeting with sub-contractors. Good central supervision
12	Work with client to minimise design changes during construction
13	Not as often as they should, however, under some contracts, say "cost plus" or alliances, there is often an incentive to do so
14	Yes, it is directly in regard to a module, process or work front, so, it is relevant to the individuals completing the tasks at management and work face
15	Depends very much on client/contractor relationships
16	It is not in contractor's interests to mitigate variations for commercial reasons
17	Orders on need basis, minimizing volume of "consumables"

By P. Rundle, January 2018

Table 6.12 Part 3 Question 33 respondent comments

No.	Respondent comments to Question 33—Part 3 waste minimization
1	Is signed onto at the start of engineering
2	Generally, under our International Organization for Standardization (ISO) 14001 Engineering Management Standard (EMS), these opportunities would be examined
3	Yes—construction methodologies and planning are for time and cost reduction maximization leading to waste reduction
4	Standard part of project plan prior to work commencement
5	This is often addressed in our corporate management plans for sustainability, not sure if it is pushed down to the contractors
6	We minimise client waste whilst preparing estimates
7	Working daily to minimise mistakes and forward thinking about each task
8	Commercial terms such as payment on surveyed installed quantities can help in this area; otherwise, in a reimbursable contract, only the client would be looking post-hardening in the case of concrete, too late
9	Contractor is obliged to submit project management plan which includes environmental management plan
10	Typically, client driven as defined in the contract
12	Again, this is standard practice throughout the detail design and project scheduling activities
13	For each project, a Waste Management Plan is prepared to address relevant elements of the building and methodology of construction
14	Some do—depends on project, government agencies moving this way
15	Australian Capital Territory requires a project-specific Waste Management Plan as part of project start-up
16	A requirement in most infra projects
17	Recycling targets included in project contract, and waste and recycling management plan in place including strategies for waste minimization
18	Categorization of potential waste and planning for segregation on-site and responsible disposal/recycling

By P. Rundle, January 2018

Table 6.13 Part 3 Question 34 respondent comments

No.	Respondent comments to Question 34—Part 3 waste minimization
1	Establish community recycling groups to identify what materials can be salvaged and re-used beneficially for the community
2	Re-use of prohibited waste products such as contaminated soil or contaminated refractory bricks within the project boundaries, suitably contained via environmental controls
3	Water reuse
4	Contractors scope to purchase/fabricate low complexity/"small" items (e.g. piping and accessories <6″)
5	Large railed collapsible shutters/formworks that were re-usable for 1.5 km of lined tunnel
6	Estimation of materials required, metal scrap bins provided, etc.
7	Redesigning to use structures that would otherwise be demolished
8	Strong recycling strategies for waste materials
9	Tonkolli first-stage A$130 million project on Green Fields Africa. Documentation was solid
10	I have seen some waste separation on domestic residential sites and commercial residential projects, especially plaster board separation
12	Develop waste management plan with recycling part
13	Known as professional project management
14	Once the contracts are let, nobody cares as long as there are a fully diverse and waste form receptacle for each form of waste on site. After that, all waste is generally dumped in the same common landfill
15	Use of quarry as spoil dump or sediment disposal area
16	Sustainability focused contractual requirements including Infrastructure Sustainability Council of Australia (ISCA) ratings and Sustainability Development Goals (SDGs) minimum scores are often effective if the requirements are well written
17	Minimizing waste is a constant on our projects, we do not always do it well and we do not have systems/processes in place to monitor our waste minimization performance
18	Built our own concrete batching plant to manage issues with concrete for cast piles and bridge decks—was largely about waste
19	Proactive policy that site waste is not buried on-site and is reused where possible
20	Unpacking deliveries off-site and then delivered periodically to construction
21	A site manager must ensure that their foremen is on top of ordering concrete during concrete pours, and volumes are correct as this is major waste product (=excess concrete) which is expensive and difficult to cart off a constrained high-rise footprint
22	Seeking resource recovery exemptions to maximise amount of material that can be reused
23	Paperless, engineering techniques

By P. Rundle, January 2018

References

Australian Constructors Association (ACA). (2017). *Australian Constructors Association Construction Outlook at June 2017*, Australian Constructors Association, viewed June to September 2017. http://www.constructors.com.au/wp-content/uploads/2017/08/Construction-Outlook-June 2017.pdf.

Charmaz, K. (2006). *Constructing grounded theory: A practical guide through qualitative analysis.* Thousand Oaks, USA: Sage.

Creswell, J. W. (1998). *Qualitative inquiry and research design: Choosing among five traditions.* Thousand Oaks, USA: Sage.

Creswell, J. W. (Ed.). (2013). *Qualitative inquiry and research design—Choosing among five approaches.* USA: Sage.

Creswell, J. W., Plano Clark, V. L., Guttman, M., & Hanson, W. (2003). Advanced mixed methods research designs. In A. Tashakkori & C. Teddlie (Eds.), *Handbook of mixed methods in social & behavioral research* (pp. 209–240). Thousand Oaks, USA: Sage.

Creswell, J. W., & Poth, C. N. (Eds.). (2017). *Qualitative inquiry and research design—Choosing among five approaches.* USA: Sage.

Dey, I. (1999). *Grounding grounded theory.* USA: Academic Press.

Faniran, O. O., & Caban, G. (1998). Minimizing waste on project construction sites. *Engineering Construction and Architectural Management, 17*(1), 57–72.

Gavilan, R. M., & Bernold, L. E. (1994). Source evaluation of solid waste in building construction. *Journal of Engineering and Management, 120*(3), 536–555.

Glaser, B., & Strauss, A. (1967). *The discovery of grounded theory.* Hawthorne, NY: Aldine Publishing Company.

Guest, G., Bunce, A., & Johnson, L. (2006). How many interviews are enough? An experiment with data saturation and variability. *Field Methods, 18*(1), 59–82.

Mason, M. (2010). Sample size and saturation in Ph.D. studies using qualitative interviews. *Forum for Qualitative Social Research, 11*(3), 1–19.

Morse, J. M. (1994). Designing funded qualitative research. In N. Denzin & Y. S. Lincoln (Eds.), *Handbook of qualitative research* (pp. 220–235). Thousand Oaks, USA: Sage.

Morse, J. M. (1995). The significance of saturation. *Journal of Qualitative Health, 5*(2), 147–149.

Nagapan, S., Rahman, I., & Asmi, A. (2012). Factors contributing to physical and non-physical waste generation in construction. *International Journal of Advances in Applied Sciences, 1*(1), 1–10.

Spivey, D. (1974). Construction solid waste. *Journal of the American Society of Civil Engineers, Construction Division, 100,* 501–506.

Strauss, A., & Corbin, J. (1998). *Basics of qualitative research: Techniques and procedures for developing grounded theory.* Thousand Oaks, USA: Sage.

Thomson, B. (2004). *Qualitative research: Grounded theory—Sample size and validity.* Ph.D. Thesis, Monash University, Melbourne, Australia.

University of New South Wales (UNSW). (2016). Construction Industry: Demolishing gender structures, UNSW report.

Yates, J. (2013). Sustainable methods for waste minimisation in construction. *Construction Innovation, 13*(3), 281–301.

Chapter 7
Findings

7.1 Demographics Findings

The pragmatic approach adopted in this textbook has helped to determine that combining sampling strategies has been the most appropriate method to meet the aims of this research and that this approach is consistent with recent developments in qualitative methods (Palinkas 2015). The primary sampling methodology utilized in this research was "purposeful" qualitative sampling, adopted to provide useful information, to learn more about the "research question" and "sub-research question" phenomenon. Purposeful sampling involves the identification and selection of a sample group of individuals and groups of individuals that are particularly knowledgeable about (and/or experienced with) the research phenomenon (Patton 2005). The Australian engineering construction industry was that target sample group. This research is unique, compared to the other studies used as benchmarks for triangulation purposes (see Faniran and Caban 1998; Nagapan et al. 2012; Sugiharto et al. 2002; Udawatta et al. 2015; Yates 2013), because this research has also used a maximum variation approach in sampling the spectrum of participants. Maximum variation sampling was chosen to allow a wide variety of participants from the engineering construction industry in different fields for the purposes, in this case, of creating a sample pool that shall be as representative as possible, in order to develop an "average" view (Creswell and Plano Clark 2011). To achieve this, a minimum sample of 50 respondents was suggested and this threshold was achieved for this research with 53 completed surveys.

The benchmark studies used for comparative purposes against this research have selectively surveyed participants from tightly banded discrete target sample groups, for example: (i) Public Works Department (Nagapan et al. 2012); (ii) senior contracting executives (Yates 2013); and (iii) bodybuilding as well as contracting companies (Faniran and Caban 1998). However, this research has differed from other construction waste studies by interviewing a broad variation of participants within the overarching umbrella of the Australian engineering construction industry, comprising of:

© Springer Nature Switzerland AG 2019
P. G. Rundle et al., *Effective Front-End Strategies to Reduce Waste
on Construction Projects*, https://doi.org/10.1007/978-3-030-12399-4_7

   I. Engaging with respondents from commercial building; residential building; heavy process construction; mining capital works; and major infrastructure Australian engineering construction sectors;

   II. Soliciting questionnaires from respondents in both the private and public sector;

  III. Obtaining survey data from contractors; builders; engineering consultants; project management specialists in scheduling, estimating, environmental control and risk; government and private sector's major capital works clients; waste facility managers; and engineering faculty academia;

  IV. Engaging with engineering construction respondents across the entire organizational chart, ranging from a cohort of 16 top management company directors from chair to board director; polling a number of highly experienced site managers, superintendents, project managers and project directors and a cadre of technical staff from the design field to the cost estimation sector;

   V. The engineering construction industry respondents who participated in this survey were highly motivated (46 per cent have worked on projects over A$2 billion), highly experienced (48 per cent have worked for over 25 years) and highly educated (45 per cent have postgraduate degrees);

  VI. The survey for this textbook has an excellent distribution of participant experience that also included under 40 respondents who, for example, are presently: (i) a site manager working on a mixed commercial high-rise development in Melbourne on his fifth similar project in succession; (ii) a project manager working in Mongolia on a heavy civil project for a top 20 Australian contractor, whose previous project with that company was as the design and construction project manager on an LNG project in Papua New Guinea; and (iii) a general superintendent on a large steel fabrication/assembly project in Western Australia;

 VII. The Australian engineering construction industry respondents who participated in this survey were polled whilst variously posted in major Australian capital cities; regional centres; remote sites; and on overseas assignments;

VIII. Engaging with the Australian peak body construction industry groups including the Master Builders Association; the Australian Construction Industry Forum; and the Australian Constructors Association;

  IX. Using two seminal models in this study, from which results from previous international studies could be readily correlated;

   X. Looking at both the sources of construction waste and the strategies to minimize site waste, whilst applying a qualitative thematic analysis approach for data synthesis to determine what common themes interrelate between construction waste sources and construction waste reduction strategies.

The construction waste sources and waste reduction strategy findings later in this chapter show that despite a broad variation of engineering construction subgroups, there is considerable homogeneity in response patterns for this research and an identifiable relationship with previous research papers used as triangulation benchmarks for data reliability purposes.

Snowball sampling was also used to obtain survey participation from the major Australian contractors by using existing respondents to recruit future participants from among their acquaintances with these major Australian Constructors Association members. The sample group built up further useful data in this manner and grew like a "rolling snowball" (Morgan 2008).

A detailed understanding of the sources of construction waste has been derived, and possible strategies to reduce site waste on Australian projects have been developed. This methodology has given voice to the often unheard people in the engineering construction industry. Again, the pragmatic approach has been used to adopt systematic sampling and generalize this research to other similar studies that have adopted the same model criteria (Faniran and Caban 1998; Nagapan et al. 2012) for validity purposes. Therefore, triangulation has been used for generalizability between one study and another (Finfgeld-Connett 2010).

7.2 Major Sources of Site Waste Findings

7.2.1 Respondent Part 2 Survey—Likert Scale Findings on Sources of Waste

Section 7.2.1 considers the research findings of the 51 completed respondent survey questionnaire answers to Likert scale Questions 12–24, inclusive, which evaluated 13 potential site waste sources on Australian construction projects based on the Faniran and Caban (1998) survey questionnaire used for this research. Acknowledgement is made of the Faniran and Caban (1998) site waste source identification model, developed upon the earlier seminal work of Spivey (1974) along with Gavilan and Bernold (1994). The Faniran and Caban (1998) model was used in this research to determine the most common sources of waste.

In addition, the Faniran and Caban (1998) approach to ranking the order of importance of the 13 potential sources of waste on the survey questionnaire, via severity index, has also been adopted in this research.

The author also opted to evaluate a ranking for the potential 13 sources of construction site waste in Questions 12–24, inclusive, of Part 2 of this research questionnaire, on account of the close scoring by using the severity index approach to rank the 12 source options. Accordingly, scores were given for each response to all 13 questions as follows: (i) "very significant" was scored a three; (ii) a "significant" answer was given a two; (iii) a response given as "of minor significance" was rated as a one; and (iv) a "not significant" reply was given a zero score. Table 7.1 provides ranking of these 13 sources of construction waste scoring each Question 12–24, inclusive, pursuant to the above.

When comparing Tables 7.1 and 7.2, the severity index ranking method and the scoring all responses method produced the same six of the seven top major sources

Table 7.1 Research survey sources of waste severity index and rank

Survey question number	Possible source of site waste	Severity index = # "very significant"/total respondents (%)	Severity index ranking
24	Lack of on-site planning and control	24	1
22	Packaging and pallet waste	20	2
12	Design and detailing errors	20	2
13	Client-initiated design changes	20	2
17	Procurement ordering and take-off errors	16	5
16	Improper materials storage	14	6
18	Poor workmanship	14	6
15	Improper material handling	10	8
14	Contractor-initiated design changes	8	9
20	Site accidents	4	10
21	Leftover off-cuts	4	10
19	Poor weather	2	12
23	Criminal waste caused by vandalism or pilfering	2	12

By P. Rundle, November 2017 (adapted from Faniran and Caban 1998)

of waste, but not in the same order. These top six major sources of site waste, not in specific order, consolidating Tables 7.1 and 7.2 are:

   I. Client-initiated changes;
  II. Design and detailing errors;
 III. Packaging and pallet waste;
 IV. Procurement ordering and take-off errors;
  V. Lack of on-site materials planning and control; and
 VI. Improper material storage.

For the sources deemed least likely to cause construction waste, the three least likely option source selections were the same but not in the same order, for Tables 7.1 and 7.2. These bottom three least likely sources of site waste, consolidating Tables 7.1 and 7.2, are:

Table 7.2 Ranking of potential sources of construction waste by scoring each response

Survey question number	Possible source of site waste	Respondent scores 0 = not significant 1 = minor significance 2 = significant 3 = very significant	Respondent score/maximum score (Maximum possible score = 153) (%)	Ranking of sources using all responses
12	Design and detailing errors	110	72	1
13	Client-initiated design changes	91	60	2
22	Packaging and pallet waste	90	59	3
24	Lack of on-site planning and control	88	58	4
15	Improper material handling	83	54	5
17	Procurement ordering and take-off errors	82	54	5
16	Improper material storage	81	53	7
18	Poor workmanship	79	51	8
21	Leftover off-cuts	75	49	9
14	Contractor-initiated design changes	58	38	10
20	Site accidents	49	32	11
19	Poor weather	47	31	12
23	Criminal waste caused by vandalism or pilfering	25	16	13

By P. Rundle, December 2017

  I. Site accidents;
 II. Criminal waste caused by pilfering and vandalism; and
III. Poor weather.

7.2.2   Comparison of Part 2 Survey Data with Other Academic Journals for Sources of Site Waste

For comparison purposes, Table 7.3 provides three sets of data from the Part 2 sources of site waste on Australian projects questionnaire results. These are from the Faniran and Caban (1998) Australian data set, from which this research adopted the same questionnaire. Also included is data from Nagapan et al. (2012), who used a similar methodology to the other two papers with the data set populated via a global literature review and using a severity index, validated by a small interview pool of participant professionals and revised the international data to fit the Malaysian engineering construction environment.

A review of the data in Table 7.3, comparing the 2017 survey results for this research and the Faniran and Caban (1998) data, with both surveys using the same questionnaire template, using an analysis of the four *most likely* sources of waste for each data set, indicates that there are three common results, which are:

  I. Design changes;
 II. Packaging and pallet waste (and other non-consumables); and
III. Design and detailing errors.

A review of the data in Table 7.3 comparing the 2017 survey results for this research and the Faniran and Caban (1998) Australian data, with both surveys using the same questionnaire template, using an analysis of the four *least likely* sources of waste for each data set, indicates that there are two common results, which are:

  I. Site accidents; and
 II. Criminal waste (from pilfering and vandalism).

The Nagapan et al. (2012) data set populated a very similar abridged Faniran and Caban (1998) site waste source identification model, after an extensive global literature review to collect this data. Nagapan et al. (2012) further developed their research, via interviews with industry professionals, for the purpose of tailoring their literature review findings to Malaysian conditions. An evaluation of the top four possible sources of site waste between this research data, the Faniran and Caban (1998) data, and the Nagapan et al. (2012) data set shown in Table 7.3, indicates that Nagapan et al. (2012) only had one common top four waste source option with the other two Australian data sets, which was "design changes".

The difference in the ranking results between this research and Faniran and Caban's (1998) top three causes of waste versus the Nagapan et al. (2012) results can possibly be explained because of the different climatic and construction maturity level between Australia and Malaysia. Malaysia, due to the vagaries of a tropical monsoon

Table 7.3 Comparison of research results with Faniran and Caban and Nagapan et al.

Research survey summary of site waste sources	Ranking	Severity index	Faniran and Caban (1998) summary of waste sources	Ranking	Severity index	Nagapan et al. (2012)	Ranking	Severity index
Lack of on-site materials planning and control	1	24	[a]Design changes	1	52.4	[a]Design changes	1	24
Packaging and pallet waste (non-consumables)	2	20	Leftover off-cuts	2	42.9	Improper material storage	2	17
Design and detailing errors	2	20	Packaging and pallet waste, etc.	3	38.1	Improper material handling	2	17
[a]Client-initiated design changes	2	20	Design and detailing errors	4	28.6	Poor weather	4	13
Procurement ordering and take-off errors	5	16	Poor weather	5	23.8	Procurement ordering and take-off errors	5	12
Improper materials storage	6	14	Improper materials handling	6	14.3	Poor workmanship	5	12
Poor workmanship	6	14	Lack of on-site materials planning and control	6	14.3	Lack of on-site materials planning and control	7	11

(continued)

Table 7.3 (continued)

Research survey summary of site waste sources	Ranking	Severity index	Faniran and Caban (1998) summary of waste sources	Ranking	Severity index	Nagapan et al. (2012)	Ranking	Severity index
Improper materials handling	8	10	Procurement ordering and take-off errors	8	9.5	Leftover off-cuts	8	8
[a]Contractor-initiated design changes	9	8	Improper materials storage	8	9.5	Poor site conditions	9	3
Site accidents	10	4	Site accidents	8	9.5	–		
Leftover off-cuts	10	4	Poor workmanship	11	4.0	–		
Poor weather	12	2	Criminal waste	12	0	–		
Criminal waste	12	2	–			–		

By P. Rundle, December 2018 (adapted from Faniran and Caban 1998)

NB[a] The research survey divided "design changes" into "client-initiated design changes" and "contractor-initiated design changes" to define this important difference environment, would place considerable emphasis on "poor weather" and "inadequate materials storage" as two key drivers in site waste sources, which Table 7.3 clearly shows.

Malaysian construction practices, which Nagapan et al. (2012) data considers, are generally of a good standard in major cities such as Kuala Lumpur. However, from the author's direct experience consulting in that country, taken across the country as a whole, Malaysia is a developing nation; therefore, materials handling and storage, with a lot of construction products, such as commonly used bagged cement for site concrete manufacture, are subject to damage and subsequently written off as site waste. Another factor that could explain a difference between the Nagapan et al. (2012) major top three waste sources results and the ranking results between this research and Faniran and Caban (1998) is the fact that Nagapan et al.'s (2012) study harvested data from an extensive global literature review to populate its sources of site waste data, duly amended to validate Malaysian conditions by interviews with only seven Malaysian industry professionals. Comparatively, this research received full completed, detailed questionnaires on sources of site waste from 51 respondents and the Faniran and Caban (1998) seminal site waste sources identification model was populated with 24 respondents (Faniran and Caban 1998, p. 184).

Nagapan et al.'s (2012) global literature review on construction waste sources resulted in them disregarding both criminal waste and site accidents as potential sources of site waste for serious consideration.

Table 7.4 provides five sets of data for comparison purposes including data from Part 2, sources of site waste on Australian projects questionnaire results of this textbook, with data synthesis using a severity index rating and a scoring all responses method. The information from the Faniran and Caban (1998) study was from an Australian data set, from which this research adopted the same questionnaire. The information from Nagapan et al. (2012), who used a similar methodology to the other two papers, populated the data set via a global literature review and, using a severity index validated by a small interview pool of participant professionals, abridged the international data to fit the Malaysian engineering construction environment. Finally, data from an Australian paper on construction waste by Sugiharto et al. (2002), which used a different method from the other two papers and this textbook is also included in Table 7.4.

The methodology in Sugiharto et al.'s (2002) academic paper differs from this textbook, Faniran and Caban's (1998) and Nagapan et al.'s (2012) research methodology approaches. However, it must be noted that the Sugiharto et al.'s (2002) paper nominates both the "design and detailing errors" option and the "design changes" option, respectively, as being the fourth and fifth main sources of site waste on Australian projects.

Udawatta et al. (2015), who provided the Australian paper evaluating site waste from a strictly human factors view, were not included in Table 7.4 comparison. This was because the author felt that this narrow, albeit valuable perspective could create bias of the consolidated view of the other papers used in Table 7.4 assessment, which

Table 7.4 Comparison of waste sources research results and three other research papers

Rating of each "source" review	Column 1	Column 2	Column 3	Column 4	Column 5
	Research survey summary of site waste sources using severity index (NB refer notes, below)	Faniran and Caban (1998) summary of waste sources using severity index (NB refer notes, below)	Research survey summary of site waste sources scoring every survey response (NB refer notes, below)	Nagapan et al. (2012) summary of site waste sources using a global literature review and sorted by severity index. Validated by interviews for Malaysian conditions (NB refer notes, below)	Sugiharto et al. (2002) Summary of site waste sources using literature review validated by survey (NB refer notes, below)
Top selection No. 1	Lack of on-site materials planning and control	Design changes	Design and detailing errors	Design changes	Poor-quality site documentation
No. 2	Packaging and pallet waste (non-consumables)	Leftover off-cuts	Client-initiated design changes	Improper materials storage	Poor weather
No. 3	Design and detailing errors	Packaging and pallet waste	Packaging and pallet waste	Improper materials handling	Unclear site drawings
No. 4	Client-initiated design changes	Design and detailing errors	Lack of on-site materials planning and control	Poor weather	Design and detailing errors
No. 5	Procurement ordering and take-off errors	Poor weather	Improper materials handling	Procurement ordering and take-off errors	Design changes
No. 6	Improper materials storage	Improper materials handling	Procurement ordering and take-off errors	Poor workmanship	Slow drawing revisions/distribution

(continued)

Table 7.4 (continued)

Rating of each "source" review	Column 1	Column 2	Column 3	Column 4	Column 5
	Research survey summary of site waste sources using severity index (NB refer notes, below)	Faniran and Caban (1998) summary of waste sources using severity index (NB refer notes, below)	Research survey summary of site waste sources scoring every survey response (NB refer notes, below)	Nagapan et al. (2012) summary of site waste sources using a global literature review and sorted by severity index. Validated by interviews for Malaysian conditions (NB refer notes, below)	Sugi...to et al. (2002) Summary of site waste sources using literature review validated by survey (NB refer notes, below)
No. 7	Poor workmanship	Lack of on-site materials planning and control	Improper material storage	Lack of on-site materials planning and control	Unclear specifications
No. 8	Improper materials handling	Procurement ordering and take-off errors	Poor workmanship	Leftover materials on site	Poor workmanship
No. 9	Contractor-initiated design changes	Improper materials storage	Leftover off-cuts	Poor site conditions	Poor coordination
No. 10	Site accidents	Site accidents	Contractor-initiated design changes	–	Lack of on-site materials planning and control
No. 11	Leftover off-cuts	Poor workmanship	Site accidents	–	Slow decision-making
No. 12	Poor weather	Criminal waste	Poor weather	–	Lack of sub-contractor skills

(continued)

Table 7.4 (continued)

	Column 1	Column 2	Column 3	Column 4	Column 5
Rating of each "source" review	Research survey summary of site waste sources using severity index (NB refer notes, below)	Faniran and Caban (1998) summary of waste sources using severity index (NB refer notes, below)	Research survey summary of site waste sources scoring every survey response (NB refer notes, below)	Nagapan et al. (2012) summary of site waste sources using a global literature review and sorted by severity index. Validated by interviews for Malaysian conditions (NB refer notes, below)	Sugiharto et al. (2002) Summary of site waste sources using literature review validated by survey (NB refer notes, below)
Lowest selection No. 13	Criminal waste	–	Criminal waste	–	Supervision too light

By P. Rundle, December 2018 (adapted from Faniran and Caban 1998; Nagapan et al. 2012; Sugiharto et al. 2002)

NB. 1—Column 1 provides a summary of the sources of site waste on Australian projects for this research using survey data calculated by severity index (see Sect. 6.5.1, "Severity Index"). The Faniran and Caban (1998) questionnaire was adopted as a template for this research

NB. 2—Column 2 provides details of the Faniran and Caban (1998) summary of the survey on sources of site waste in Australia, calculated by severity index, and was used as a template for this research. Survey participants were drawn from both building and contracting firms

NB. 3—Column 3 provides details for the sources of site waste on Australian projects summary from this research, using the same raw data as Column 1, but calculated by apportioning a score to each and every respondent answer and weighting the results. The Faniran and Caban (1998) questionnaire was adopted as a template for this research

NB. 4—Column 4 provides details for the sources of site waste internationally by Malaysian academics Nagapan et al. (2012) via a global literature review, ranked via severity index and validated for Malaysia by a limited survey of local construction professionals. Columns 1, 2, 3 and 4 have used the same survey questionnaire approach, and therefore, results can readily be compared. Acknowledgement is made of this to Faniran and Caban (1998), whose boilerplate questionnaire was adopted by the author for this research survey and to Nagapan et al.'s (2012) work in this area, building on Spivey (1974); Bossink and Brouwers (1996); Gavilan and Bernold (1994)

NB. 5—Sugiharto et al. (2002), summarized in Column 5 of Table 7.4, adopted a different approach to their evaluation of potential sources of site waste on Australian projects. Through an extensive literature review and through pilot studies, Sugiharto et al. developed an independent set of 31 variables that they used to undertake a survey questionnaire with contractors to rank potential sources of site waste. Six of the top seven sources were design and documentation related variables. Accordingly, though the data used in Column 5 is valuable and used in this overall assessment, the results can be seen as an "outlier" to Column 5 results, which basically used the same assessment framework was an "open" assessment of factors. Notwithstanding this, the Udawatta et al. (2015) findings have been considered later in this report, during Chap. 8, "Discussion".

An evaluation of the lowest ranked four sources of waste options in Table 7.4, "Comparison of Waste Sources Research Results and Three Other Research Papers", clearly indicates that the least likely sources of waste are "criminal waste" and "site accidents". Another observation from a review of Table 7.4 is the site waste source identification model, abridged by the author in this research to account for "client-initiated design changes" and "client-initiated design changes", rather than the seminal Faniran and Caban (1998) model that simply considers "design changes". Table 7.4 clearly shows a marked difference in this potential source of site waste from "design changes". This research respondent survey data synthesis rates "client-initiated design changes" as the second most important source of site waste. However, respondents for this research give "contractor-initiated design changes" a very low rating at tenth place. All other three papers evaluated in Table 7.4, namely Faniran and Caban (1998); Nagapan et al. (2012) and Sugiharto et al. (2002), give "design changes" a high rating as an important source of site waste. *Prima facie*, there is a case to sub-categorize future research using the Faniran and Caban (1998) site waste source identification model to more accurately address the root cause of a very important source of waste, as outlined by industry professionals in all four of the aforementioned studies, including this textbook.

Table 7.4 also indicates that the "poor weather" option presents a primary difference between this textbook and all three papers considered in this Sect. 7.2.2 on sources of site waste (Faniran and Caban 1998; Nagapan et al. 2012; Sugiharto et al. 2002). Although an evaluation of the research data from respondent surveys for this book rated "poor weather" as one of the lowest potential sources of site waste, the three other papers all rated "poor weather" as one of the top five causes of waste. This anomaly needs to be the subject of further research, to provide an explanation, albeit empirically.

7.2.3 *Summary Findings on Part 2 Questions 12–24, Sources of Site Waste*

This section summarizes research findings of respondent survey answers to Likert scale Questions 12–24, inclusive, which evaluated 13 potential site waste sources on Australian construction projects based on the Faniran and Caban (1998) survey questionnaire used for this research. The respondents were asked to evaluate the 13 potential sources of waste as being either: (i) "not significant"; (ii) "of minor significance"; (iii) "significant"; or (iv) "very significant". Quite clearly, the evaluation synthesis of the available research data indicates that the top six causes of site waste are:

I. Client-initiated changes;
II. Design and detailing errors;

III. Packaging and pallet waste;
IV. Procurement ordering and take-off errors;
 V. Lack of on-site materials planning and control; and
VI. Improper material storage.

A comparison with the Faniran and Caban (1998) Australian data, from which this research questionnaire was taken, shows that both this research and the Faniran and Caban paper concur that the top three sources of waste are:

 I. Design changes;
 II. Packaging and pallet waste (and other non-consumables); and
III. Design and detailing errors.

These are the top three rankings for this data set as well, but not in the same order.

A comparison between the Nagapan et al's. (2012) Malaysian data and this text-book's data, along with the Faniran and Caban (1998) data, indicates that there is consensus between all data sets that "design changes" are the highest ranked source of site waste. A similar comparison using the research data set, along with the Faniran and Caban (1998) research as a benchmark against the Sugiharto et al. (2002) data set, shows that the "design and detailing error" option and the "design changes" option are among the top five sources of site waste, after a review using industry participant surveys.

In summary, regarding the major sources of site construction waste, both the research data set comprising a synthesis of 51 respondent replies to 13 questions on the sources of waste and the consolidated five sets of data from Table 7.5 indicate that the six main sources of construction waste are:

 I. Design and detailing errors;
 II. Client-initiated changes;
III. Packaging and pallet waste;
IV. Procurement ordering and take-off errors;
 V. Lack of on-site materials planning and control; and
VI. Improper material storage.

Regarding "criminal waste" and "site accidents" as potential causes of waste, the Nagapan et al. (2012) assessment, which aligns with the Faniran and Caban (1998) model used in this research, did not even consider either "criminal waste" or "site accidents" after their comprehensive global literature review suggested that these two factors were considered to be potentially minor sources of site waste. Sugiharto et al. (2002), using a different model than this research, also entirely discounted "criminal waste" and "site accidents" as potential site waste sources. Though the author has directly experienced large-scale pilfering of equipment and materials on a major Australia construction project and malicious and continual damage on a large project in Papua New Guinea, *prima facie*, there is a case for the judicious inclusion of the "criminal waste through theft and vandalism option" on future site waste survey questionnaires, only if the researcher on future studies believes that this

Table 7.5 Top seven sources of waste from synthesis of this research and three other papers

Consolidated evaluation of the five assessments of sources of construction waste—top seven sources Source description	Consolidated evaluation of the five assessments of sources of construction waste—top seven sources Ranking	Consolidated evaluation of the five assessments of sources of construction waste—top seven sources Number of times option was selected for top seven choices for each assessment
Design and detailing errors	Top selection No. 1	4
Design changes	No. 3	4
Packaging and pallet waste	No. 3	3
Procurement ordering and take-off errors	No. 3	3
Lack of on-site materials planning and control	No. 3	3
Improper materials storage	No. 3	3
Poor weather	No. 3	3

By P. Rundle, January 2018

option is likely valid. As for "site accidents" being a source of physical site waste, this research has shown, along with results from the three other papers used for data triangulation (Faniran and Caban 1998; Nagapan et al. 2012; Sugiharto et al. 2002), that there is little or no propensity for "site accidents" being a significant source of construction site waste.

In summary, it is obvious that the "criminal waste" from pilfering and vandalism and the "site accident" sources of site waste options were not considered as having a high causation potential for both this research as noted in Sect. 7.2.1 of this report, or in any data from the three other papers on sources of waste used in Sect. 7.2.2 for comparative purposes.

Prima facie, there is a case to sub-categorize future research using the Faniran and Caban (1998) site waste source identification model option, which nominates "design changes" as a potential source of site waste. The author posits that the model needs to account for two new waste sources instead, one option entitled "client-initiated design changes" and the other options being "contractor-initiated design changes", to more accurately address the root cause of a very important source of waste, as outlined by industry professionals in all four of the aforementioned studies, including this textbook.

There was one anomaly in the results between this book and the other three papers considered in this Sect. 7.2.3 on sources of site waste (Faniran and Caban 1998; Nagapan et al. 2012; Sugiharto et al. 2002). Although an evaluation of this research data of respondent surveys for this book rated "poor weather" as one of the lowest potential sources of site waste, the three other papers all rated "poor weather" as one of the top five causes of waste. "Poor weather", as a cause of construction waste, will be further discussed in Chap. 10, "Recommendations", of this report as an area that should be researched further.

7.2.4 Thematic Analysis of Part 2 Sources of Site Waste Using Coding/Themes/Sub-themes

This section uses a thematic analysis research approach to dissect the information obtained in Part 2 on the sources of construction site waste survey questionnaire and, more particularly, Question 25, which asks respondents to comment on other sources of site waste, over and above the 13 options noted in Questions 12–24 of this Part 2 questionnaire.

The survey questionnaire used to provide respondent data to run Faniran and Caban's (1998) site waste source identification model did not seek information from respondents. However, Question 25 of the survey questionnaire, introduced as an adaption by the author to the Faniran and Caban (1998) model, requested respondents to provide other possible sources of site waste, over and above, those nominated in Question 12 to Question 24, inclusive. The author selected a thematic analysis as the most appropriate research tool to synthesize respondent comments on construction site waste sources.

Section 7.2.1 used a Likert scale to collate respondent's opinions on 13 possible sources of site waste contained in Questions 12–24, inclusive, in Part 2 of this research survey. Both the survey questionnaire template and the severity index ranking approach, to analyse the data set, were adopted from an earlier paper on Australian project site waste by Faniran and Caban (1998). Though the Likert scale was used to collect respondent opinions, Faniran and Caban (1998) adapted this methodology for use in a qualitative analysis on construction waste sources.

Thematic analysis focuses on evaluating the themes captured within the data (Creswell et al. 2003). Coding is the main process for preparing and developing themes from the raw data (Creswell et al. 2003). Thematic analysis is consistent with the phenomenology approach adopted for this research, which focuses on the human subjective experience that has emphasized participant's perceptions, opinions and experiences on the topic (Creswell and Poth 2017). The thematic analysis approach is also recommended as a suitable research tool in the qualitative pragmatic framework used in this research (see "validity" in the "standards" column of Table 5.2, entitled, "Framework for a Pragmatic Qualitative Research Approach", of Sect. 5.5, "Framework for a Pragmatic Approach").

Themes are patterns across data sets that are important to the description of a phenomenon and are associated with the specific research question (Creswell et al. 2003). Sub-themes are discrete patterns, identified within a theme (Guest 2012).

Nagapan et al. (2012, p. 5, Table 3) carried out a global literature review of the main causes of construction waste and tabulated these findings into themes (categories) and codes (factors). Acknowledgement is made of the Nagapan et al. (2012) work, building on Spivey (1974); Gavilan and Bernold (1994); and Faniran and Caban (1998) from whom Nagapan et al. (2012) used an abridged site waste source identification model. This textbook used the Nagapan et al. (2012) findings, as appropriate, to identify codes and primary themes as a starting point to develop the primary themes. The author developed sub-themes, categorizing patterns within themes, independently of these seminal papers. It should be noted that there are several themes and many codes contained herein, in Table 7.7, developed during synthesis of this research data, which do not appear in the Nagapan et al. paper (2012, p. 5, Table 3).

7.2.4.1 Summary Table of Thematic Analysis of Sources of Waste, as Per Respondent Survey Part 2, Question 25

This section undertakes a thematic analysis of respondent comments provided from survey questionnaire Part 2, Question 25, which states: "*list any other sources in your opinion that cause site construction waste*". These responses cover the respondent's opinions of site waste sources over and above the 13 potential sources noted in Questions 12–24, inclusive, of Part 2 of the survey. The rich pool of data leads the author to suggest that future research on site waste sources amends the Faniran and Caban (1998) model questionnaire with this extra question and uses thematic analysis as the appropriate tool to interpret comments (Table 7.6).

7.2.4.2 Key Points from the Author's Reflexivity Journal on Sources of Site Waste from Question 25 Respondent Comments for Thematic Analysis Process

Table 7.7 provides the key factors from the author's reflexivity journal that progressively logged reflections on emerging patterns, themes and concepts during the data analysis for the nominated respondent comments. These reflections were progressively recorded from the coding stage, right through to the logging of any changes to themes/sub-themes during the review phase (Saldana 2009).

7.2.4.3 Themes Developed, as Per Question 25 Respondent Comments on Sources of Waste

This section is a thematic analysis developed from a synthesis of respondent data, as per survey questionnaire Part 2, Question 25, which states: "*list any other sources in your opinion that cause site construction waste*". These responses cover the respondents' opinions of site waste sources over and above the 13 potential sources noted in Questions 12–24, inclusive, in Part 2 of the survey. Note that several of the respondents had more than one comment when addressing Question 25.

Table 7.6 Development of coding/themes/sub-themes from respondent sources of waste data

No.	Respondent opinions on other possible sources of waste, questionnaire Part 2, Question 25	Coding	Theme	Sub-theme
1	Inaccurate procurement processes that cause site construction waste	Procurement errors	Procurement	Quality process shortfall
2	Generally good planning, procurement and delivery schedules	Procurement processes	Procurement	–
3	Packaging and temporary formwork(s) of poor quality	Packaging/poor formwork	Procurement/material	Quality process shortfall
4	Overengineering by designers and lack of optimization and lack of value engineering	Overdesign/need good early design processes	Design	FEED (=front-end engineering design)/quality process shortfall/client responsibility
5	Incorrect classification of waste (e.g. classifying something to be sent to landfill when it actually could have been re-used or recycled); lack of understanding about local waste classification and guidelines; not knowing what materials can be recycled/re-used	Incorrect waste-type classification	Management	Recycling resources
6	Poor segregation (of construction waste)	Incorrect waste-type classification	Management	Recycling resources
7	Vendors	Sub-standard material quality	Material	Quality process shortfall
8	Poor housekeeping	Poor site clean-up	Site conditions	Quality process shortfall
9	All major causes covered in above Question 12 to Question 24 complete	N/A	N/A	–
10	Poor planning	Poor planning	Management	Quality process shortfall

(continued)

Table 7.6   (continued)

No.	Respondent opinions on other possible sources of waste, questionnaire Part 2, Question 26	Coding	Theme	Sub-theme
11	Lack of proper supervision	Lack supervision	Management	
12	Ability to recycle certain plastics and concrete surplus often results in waste to landfill	Recycled waste products dumped in landfill	Management	–
13	Logistics of concrete deliveries	Material logistics	Procurement	–
14	Poor overall management and its enforcement by key site personnel	Poor supervision	Management	Quality process shortfall
15	Improve client review and approval of management plans and design submissions	Poor client review process	Management	Client responsibility
16	Inadequate clean-up of work areas on a continuous basis	Poor site clean-up	Site conditions	Quality process shortfall
17	Client requirement for samples and prototypes	Early design involvement	Design	FEED
18	Insufficient storage facilities, leading to spoiling of materials (e.g. bags of cement left in the open)	Insufficient materials storage	Procurement	Quality process shortfall
19	Site clean-up at end of project	Poor site clean-up	Site conditions	Quality process shortfall
20	Oversights in planning leading to re-work and re-design, in order to secure relevant approvals before start of project, or to progress on new stage/extension	Poor planning	Management	Quality process shortfall
21	Work culture not valuing materials	Poor work culture	Management/workers	Quality process shortfall/work culture
22	No consideration of construction waste minimization as a primary design feature	Need good early design	Design	FEED

(continued)

Table 7.6 (continued)

No.	Respondent opinions on other possible sources of waste, questionnaire Part 2, Question 25	Coding	Theme	Sub-theme
23	Individual trades not being held accountable for the waste they generate	Poor work culture by tradespeople	Workers	Work culture
24	Oversupplied materials	Ordering errors	Procurement	Quality process shortfall
25	Engineers and architects	Need good design	Design	Work culture
26	Incorrect estimation of material	Ordering errors	Procurement	Quality process shortfall
27	Defective materials or workmanship	Poor material/poor workmanship	Material/workers	Quality process shortfall
28	Poor personal hygiene practices by sub-contractors/blue collar workers and poor clean-up as project progresses	Poor work culture by workers	Workers	Quality process shortfall
29	Lack of attention to early design effort	Early design involvement	Design	FEED
30	Overordering and poor handling	Poor ordering/poor handling	Procurement	Quality process shortfall
31	Lack of suitable outlets; space limitations on site [e.g. space to segregate and sort, timing of waste production versus availability of recyclers (inert material)]	Inadequate recycling resources	Management	Recycling resources
32	10 per cent wastage adds up to a very significant factor in construction wastage	Ten per cent wastage allowance too high	Management	–
33	Poor workmanship causing breakout work; lack of concern by section gangs and their engineers to wastage and primary materials	Poor workmanship/poor management	Workers/management	Work culture
34	Poor supervision of materials control, at installation/build point	Poor materials control	Procurement	Quality process shortfall

(continued)

Table 7.6   (continued)

No.	Respondent opinions on other possible sources of waste, questionnaire Part 2, Question 2?	Coding	Theme	Sub-theme
35	Mostly poor quality of materials and/or equipment, from China and other Asian locations	Poor material quality	Material	Quality process shortfall
36	Not engaging (client) operations early in the design	Early design involvement	Design	FEED/client responsibility
37	Insufficient waste collection and waste (storage) facilities. Uncovered skips	Insufficient waste storage facilities	Management	Recycling resources
38	Lack of planning and actively promoting recycling and re-use of "waste" materials for "beneficial" use of local communities	Waste storage facilities	Management	–
39	Engineering designers and architects do not design to mitigate waste	Early design involvement	Design	FEED
40	Contractors pass on material waste costs to clients. Ultimately, clients are a source of waste by not configuring a mandatory waste reduction plan in contract as part of its tender evaluation	Poor contractor attitude	Management	Work culture/client responsibility
41	Not specifying recyclable materials, such as filters, etc. (in the permanent works) during the design and maintainable baseline development	Good design specifications	Design	FEED
42	No segregation when recycling	Inadequate recycling resources	Management	Recycling resources
43	Allowing "overcontingency" when ordering	Ordering errors	Procurement	Quality process shortfall

By P. Rundle, January 2018

Table 7.7 [a]Extracts from the author's reflexivity journal of key points for respondent comments

No.[b]	Respondent opinions on possible causes of waste over and above questionnaire Part 2, Questions 12 to 24, inclusive	Key points from the author's reflexivity journal whilst developing codes/themes/sub-themes
4	Overengineering by designers (lack of optimization and lack of value engineering)	Client responsibility to ensure performance requirements specification to consultants, specify value engineering workshop
5	Incorrect classification of waste (e.g. classifying something to be sent to landfill when it actually could have been re-used or recycled); lack of understanding about local waste classification and guidelines; not knowing what materials can be recycled/re-used	Author's initial response to respondents discussions on recycling was either: (i) those respondents failed to understand that survey pertained to "front-end" and not "back-end" waste minimization, as was stated in the questionnaire; (ii) author never made this point clearly enough; or (iii) respondents are in such a rush to complete surveys that they often do not read the detail and author should bear this in mind for future research projects. However, as coding was occurring and themes emerged, author reflected that providing adequate personnel resources to classify waste for recycling was actually a front-end waste reduction initiative
18	Insufficient storage facilities, leading to spoiling of materials (e.g. bags of cement left in the open)	Nagapan et al. (2012, Table 3), was an excellent preliminary guide to reflect on code and major theme development as a starting point. This paper's detailed literature review developed 81 key factors on sources of waste as using the seminal Spivey (1974) model for site waste classification, as a framework. Nagapan et al. (2012) then bracketed these 81 factors (codes) into groups (themes) to develop a site waste sources matrix. Nagapan et al. proceeded to validate this matrix via respondent survey/interviews. The advantage for this research regarding data trustworthiness is that the Faniran and Caban (1998) paper survey questionnaire was used in this research and can be related straight back to Nagapan et al. (2012) as both papers used the Spivey (2014) model as the research framework. However, in the code example shown here, Nagapan et al. (2012) have classified this as being a "handling" group classification, where to the author quite clearly, this is a "procurement" theme. This is borne out by Faniran and Caban (1998) which also suggested that material storage was a procurement issue for material waste

(continued)

Table 7.7   (continued)

No.[b]	Respondent opinions on possible causes of waste over and above questionnaire Part 2, Questions 12 to 24, inclusive	Key points from the author's reflexivity journal whilst developing codes/themes/sub-themes
4 4	Poor workmanship causing breakout work; lack of concern by section engineers and gangs, to wastage and primary materials	This is an example of an issue that is immediately recognizable in the Spivey (1974) model used in this book as a framework for data classification (poor workmanship), but also points to a management issue and an underlying work culture issue
36	Not engaging (client) operations early in the design	The client needs to ensure that its performance requirement specifications to engineering and construction personnel cites this special requirement, but all parties must agree to the extent of client operations personnel involvement to avoid delay to the design/build/commission process
39	Engineering designers and architects do not design to mitigate waste	This is an example of an immediately apparent "design"
N/A	General remarks	When determining themes, which broadly will be readily apparent, but more so, sub-themes, which will require a synthesis of data to determine patterns across codes/themes, be aware of confirmation bias that could "slant" author's approach. Concentrate on academic models/theories to underpin data synthesis

By P. Rundle, February 18

[a]NB—Table 7.7 captures only the author's primary reflections of relevant respondent comments

[b]NB—The number assigned to items in Table 6.5 corresponds to the primary item number found in Table 6.4, Sect. 7.2

"Management" Theme

As can be noted in Table 7.8, the "management" theme is the major factor impacting site waste sources. The "management" theme was developed from coding of participant remarks and factors identified, including such issues as poor planning; lack of competent field supervision, including underperforming section engineers; poor work culture exhibited by workers and technical staff; and an industry accepted high material waste factor of 10 per cent. There is a need for project management field teams to be provided with the latest Information Communications Technology (ICT) tools, as well as better supervision to effectively allocate recycling resources and to arrange for prudent waste identification/segregation. Client management must take responsibility for adequately scoping the construction waste mitigation plans in all of its contractual arrangements, insisting on consultant waste reduction design plans in tenders and ensuring that their contract documents allow for an equitable appor-

Table 7.8 Site waste themes developed from participant survey data on waste sources

Theme No.	Theme	Number of times this theme was recorded against respondent comment	Remarks
1	Management	16	
2	Procurement	10	
3	Design	9	
4	Workers	5	
5	Materials	5	
6	Site conditions	3	This theme context refers to regular clean-up of site waste
7	N/A	1	Unable to determine meaning of respondent comments

By P. Rundle, February 2018

tionment of waste management responsibility shared with contractors. Contractor management must also accept its fair share of waste management risk and build a work culture of construction waste reduction with its manual and technical staff, rather than simply passing cost over-run risk on to the owner.

In summary, the survey respondent sample pool, which actually comprises members of the engineering construction management profession in various fields including engineering, contracting, field supervision, clients, technical and project management fields, has overwhelmingly nominated the "management" cadre as the most important theme related to site waste sources.

"Procurement" Theme

Table 7.8 indicates that the "procurement" theme was identified as the next most critical theme impacting site waste sources. Codes identified procurement errors in areas such as poor material control, including overordering, incorrect material take-offs and other ordering errors; poor logistics, including improper material handling and poor storage facilities; and improper handling of materials at the work face. The predominant re-occurring code for the "procurement" theme was the requirement for an integrated materials supply chain as a process to drive project materials flow as an important potential waste source control process.

"Design" Theme

Table 7.8 indicates that the "design" theme was the next most important theme in creating possible sources of site waste. The predominant code for this theme was a requirement to undertake adequate "front-end engineering design" (FEED) to control construction waste. Other codings referred to a need to design, where appropriate, a system that allowed steel structures to be modularized off-site; the implementation of value engineering workshops; and ensuring that the design specification accounts for construction waste mitigation.

Table 7.9 Site waste sub-themes developed from participant survey data

Sub-theme No.	Sub-theme description	Number of times this sub-theme was recorded against respondent comment
1	Quality process shortfall	18
2	Front-end engineering design (FEED)	7
3	Work culture	4
4	Recycling resources	4
5	Client responsibility	4

By P. Rundle, February 2018

"Workers" Theme

Table 7.8 shows that there were around half as many respondent comments for the "workers" theme as for the "design" theme. The main respondent codes concerned improving worker culture to account for wasting materials; addressing poor workmanship; and developing trades person competencies.

"Materials" Theme

Defective materials were the main code identified by respondents for this "materials" theme.

"Site Conditions" Theme

All three codes for this "site conditions" theme relate to the necessity to continually remove construction waste from the site for the duration of the project, right up to final demobilization clean-up.

7.2.4.4   Sub-themes Developed, as Per Question 25 Respondent Comments on Sources of Site Waste

A review of patterns, from both within and across themes, has provided a list of sub-themes described in Table 6.4 of this textbook and summarized in Table 7.9. The following is a synthesis on each sub-theme.

"Quality Process Shortfall"

Table 7.9 clearly shows that the "quality process shortfall" theme is the major sub-theme factor impacting site waste sources. The "quality process shortfall" sub-theme encompasses project management processes that have not been followed and/or project management work activities that require a process to be prepared and implemented. This sub-theme has been prepared using the rationale that if a quality assurance audit was undertaken, then a shortfall (i.e. non-compliance) would uncover the waste source causation noted by the respondent in their remarks on construction waste source. For example, defective material quality, poor material handling, incorrect material storage, overordering, incorrect ordering, poor-quality formwork and sub-standard workmanship are all quality assurance shortfalls (i.e. process non-compliances) in the project management system. Sub-standard supervisory staff and poor project planning are considered quality non-compliances.

"FEED"

"FEED", the next most critical sub-theme, is a specific design process that addresses the criteria under which engineering design must be executed from engineering inception, with a view to minimizing site waste, and must be a mandatory contract requirement (Yates 2013). Using this approach, the consultant can maximize the opportunities to reduce site waste. Using such designs that allow modularization of process equipment; pre-assembly of components; off-site fabrication of work elements such as pipework; and precast concrete element fabrication, all promote off-site assembly in a controlled environment, which mitigates both site waste sources and waste volume.

"Work Culture"

The three next most important sub-themes in this research as per Table 7.9, are "work culture", "recycling resources" and "client responsibility". First, the "work culture" sub-theme shall be discussed. The synthesis of respondent data on sources of site waste indicates that the perceived culture of workers, supervisors, management, consultants and clients not being interested in the reducing root cause(s) of construction waste on Australian projects needs to be altered by the same process that has seen the adaptation of recycling as the accepted work culture for site waste process reduction. This cultural change was introduced and has been maintained via a tripartite approach comprising of government-run community and industry engagement campaigns; government regulations; and finally, enforcement by punitive legislation, to make recycling of site waste products the regular practice internationally, particularly in developed world economies (Chandler 2016; Hickey 2015; Ritchie 2016).

"Recycling Resources"

Table 7.9 indicates that the next highest sub-theme equal in respondent's perception with the "work culture" and "client responsibility" sub-themes, is "recycling resources". As noted in the author's reflexivity journal in Table 7.7 of this textbook the author's initial perception was that respondents addressing recycling processes had failed to properly read their pro forma survey documents, which precluded recycling because this research studied upstream waste reduction rather than downstream waste handling. However, an analysis of comments from several respondents showed that recycling efforts were being hampered at the front end of the construction waste cycle by (i) a lack of experienced personnel to grade construction waste products for potential recycling or re-use; (ii) an inadequate number of recycling bins, strategically placed around the site; (iii) insufficient personnel to recycle into bins on site; and (iv) the lack of a laydown area resource to be allocated for recycling. It is regularly recognized that on a commercial high-rise central business district project, there would be limited capacity for this.

In summary, the provision of these additional resources to assist in recycling, throughout the duration of a project, would definitely aid in front-end site waste minimization.

"Client Responsibility"

The "client responsibility" sub-theme was equal to the "recycling resources" sub-theme, as shown in Table 7.9. A mature client has a set of standards, plans, drawings, technical specifications and contract terms and conditions that describe exactly what the engineering consultant, contractor and vendor must provide (BHP Billiton 2007). Respondent comments in Table 6.1 indicate that the "client responsibility" sub-theme encompasses: (i) limits on the responsibility between client and contractor for handling waste; (ii) value engineering and optimization workshops that aim at evaluating waste reduction as part of a review; (iii) sustainable use of materials; and (iv) front-end engineering design that accounts for waste reduction. It remains with the client to provide a set of tender and, subsequently, contractual documents that account for all of these waste minimization processes for execution by the contractor, engineering consultant and vendor.

7.2.4.5 Summary Table of an Abridged Thematic Analysis of Sources of Waste, as Per Respondent Survey Part 2, Questions 12–24, Inclusive

Part 2 survey, Questions 12–24, asked respondents to indicate the relative significance of construction site waste sources by indicating if the source was "very significant"; "significant"; "of minor significance"; or "not significant". In accordance with the Faniran and Caban (1998) model (from which Part 2 questionnaire for this research was adopted and from which each of the 13 construction site waste sources was identified for this survey), a severity index was determined by calculating the percentage of respondents giving the response "very significant" and the 13 site waste sources were ranked on this basis.

For the purposes of maximizing utilization of this data, the author has endeavoured to use an abridged and simplified form of thematic analysis to evaluate the top six sources of waste, using the Likert scale data of 51 industry respondent comments obtained in Questions 12–24. This novel approach was taken for the express purpose of developing themes and sub-themes, to expand data pool utilization and add to the body of work. Even though the rich pool of respondent comments was not provided for these 13 questions on waste sources, the condensed view of themes and sub-themes is possible because of the nature of how the questions pertaining to the 13 primary sources of waste, in Faniran and Caban's (1998) site waste identification model, have been framed.

Building on the earlier work of Faniran and Caban (1998), the author has also used an abridged thematic analysis to evaluate the results from the research findings in Sect. 7.2.4. The author argues that it is reasonable, given the pragmatic framework underpinning this research, to analyse the results for Part 2, Questions 12–24, inclusive, 13 possible sources of construction waste, even though the responses were provided via a Likert scale, by a simple abridged form of thematic analysis. Each of the 13 sources of waste addressed in Part 2, Questions 12–24, inclusive, survey questionnaire for this research, corresponds to a code that has been bracketed into a theme by Nagapan et al. (2012, p. 5, Table 3). This is possible because Nagapan et al. (2012) followed the seminal approach established by Spivey (1974) and developed by later academic papers, whilst developing codes and corresponding primary themes and bracketing codes. The sub-themes were synthesized by the author. The following six main options have been noted as the major sources of construction site waste from an analysis of survey data for this research.

   I. Client-initiated changes;
  II. Design and detail errors;
 III. Packaging and pallet waste;
 IV. Procurement ordering and take-off errors;
  V. Lack of on-site materials planning and control; and
 VI. Improper material storage.

These six potential sources were also identified as codes in the thematic synthesis of Part 2, Question 25, data, which comprised of respondent comments on other possible sources of site waste, over and above the 13 possible sources described in Part 2, Questions 12–24. From these codes, corresponding themes and sub-themes were obtained, from the thematic analysis of Part 2, Question 25 data, which comprised of respondent comments on other possible sources of site waste, over and above the 13 possible sources described in Part 2, Questions 12–24.

These results, summarized in Table 7.10, indicate that the main theme was "procurement" across the six major potential sources of waste, determined via a survey questionnaire of 51 respondents covering Questions 12–24 with Likert scale responses ("not significant"; "of minor significance"; "significant"; and "very significant"). The next two themes of equal significance were "design" and "management". However, the primary thematic analysis undertaken using Part 2, Question 25 respondent comments on potential other site wastes summarized in Table 7.8, found that the major theme was "management", with "procurement" being only the second most common theme, albeit clearly second.

Table 7.10 shows that "quality assurance shortfalls" were the predominant sub-theme, which was followed equally by "client responsibility" and "innovative packaging solutions". The primary sub-theme analysis undertaken using Part 2, Question 25 respondent comments on potential other site waste, detailed in Table 7.9, found that the major sub-theme was also "quality assurance shortfalls".

7.2.4.6 Summary of Thematic Analysis of Theme and Sub-theme Findings for Part 2—Sources of Site Waste Survey

Themes

The primary source of rich respondent data was provided by Part 2, Question 25, of this research survey questionnaire requesting participants to identify "other" possible sources of site waste over and above the 13 options cited in the Faniran and Caban (1998) site waste identification model used in this research. As noted elsewhere the

Table 7.10 Development of coding/themes/sub-themes from respondent sources of waste data

[a]No.	Coding	Theme	Sub-theme
1	Client-initiated changes	Management	Client responsibility
2	Design and detail error	Design	Quality process shortfall
3	Packaging and pallet waste	Procurement	Innovative packaging solutions
4	Procurement ordering and take-off errors	Procurement	Quality process shortfall
5	Lack of on-site materials planning and control	Procurement	Quality process shortfall
6	Improper material storage	Procurement	Quality process shortfall

By P. Rundle, February 2018

[a]NB—There are no respondent comments in Table 7.10 because an abridged simplified thematic analysis was employed for this Sect. 7.2.4.6 only. Comparatively, the six waste sources in Table 7.6 were identified by a code and then bracketed into a theme derived from similar codes in the body of work elsewhere in this book, as per the thematic analysis on Question 25 covering respondent comments on sources of waste

Table 7.11 Consolidated summary of themes for Part 2—sources of site waste

[a]Theme No.	Theme	Number of times this theme was recorded against respondent comments
1	Management	19
2	Procurement	15
3	Design	9
4	Workers	5

By P. Rundle, February 2018

[a]NB—The "site conditions" theme was consolidated into the "management" and "materials" themes and rolled into the "procurement" theme

author has followed theme development using Nagapan et al.'s paper (2012, p. 5, Table 3) as appropriate. However, upon reflection of this particular theme development approach, the "site conditions" theme, comprising solely of site clean-up, is in fact a field management responsibility. Similarly, a review of the permanent material product quality compliance, as per the "materials" theme, can be readily associated as a "procurement" theme function (Table 7.11).

An abridged thematic analysis (refer to Table 7.10) determined that the "procurement" theme was the main theme across the top six major potential sources of waste, developed via a survey questionnaire of 51 respondents covering Questions 12–24 with Likert scale responses ("not significant"; "of minor significance"; "significant"; and "very significant") and without the benefit of respondent comments. The next

Table 7.12 Sub-themes for Part 2 sources of site waste

Sub-theme No.	Sub-theme description	Number of times this sub-theme was recorded against respondent comment
1	Quality process shortfall	18
2	Front-end engineering design (FEED)	7
3	Work culture	4
4	Recycling resources	4
5	Client responsibility	4

By P. Rundle, February 2018

two themes of equal significance were "design" and "management". However, the primary thematic analysis undertaken using Part 2, Question 25, respondent comments on potential other site waste, summarized in Table 7.8, found that the major theme was "management", with "procurement" being only the second most common theme, albeit clearly second.

Accordingly, the key themes that this research points to as being the most relevant to construction site waste sources are, in descending order:

  I.   Management;
 II.   Procurement;
III.   Design; and
IV.   Workers.

In the following Chap. 8, "Discussion", a detailed appraisal has been carried out of these four major themes, from which conclusions have been drawn and recommendations made for this research surrounding sources of construction site waste.

Sub-themes

Table 7.12 summarizes, in descending order, the main sub-themes developed from respondent comments to Part 2, Question 25, which asked for the participant's recommendations for alternative sources of waste to the 13 options nominated in Part 2, Questions 12–24, sources of site waste survey.

In summary, the main sub-theme for sources of site waste is unambiguously "quality process shortfalls", followed by "front-end engineering design" and then a three-way tie for third, for the "work culture", "recycling resources" and "client responsibility" options. An abridged simplistic thematic analysis (refer to Table 7.10) determined that the "procurement" theme was the main theme across the top six

Table 7.13 Consolidated sub-themes for Part 2 sources of site waste	Sub-theme No.	Sub-theme description	Number of times this sub-theme was recorded against respondent comment
	1	Quality process shortfall	77
	2	Front-end engineering design (FEED)	7
	3	Work culture	4
	4	Client responsibility	4

By P. Rundle, February 2018

major potential sources of waste, developed via a survey questionnaire of 51 respondents covering Questions 12–24 with Likert scale responses ("not significant"; "of minor significance"; "significant"; and "very significant") and without the benefit of respondent comments. Table 7.10 shows that "quality assurance shortfalls" was the predominant sub-theme, followed equally by "client responsibility" and "innovative packaging solutions". The primary sub-theme analysis undertaken using Part 2, Question 25, respondent comments on potential other sources of site waste, detailed in Table 7.9, found that the major sub-theme was also "quality assurance shortfalls" (Table 7.13).

It can be persuasively argued that the lack of adequate recycling resources, whether they be adequate personnel for waste sorting or enough recycling bins, can be construed as a failure of a quality process for the site recycling standard operating process.

In summary, the four sub-themes involved in site waste sources in descending order are:

I. Quality process shortfall;
II. Front-end engineering design (FEED);
III. Work culture; and
IV. Client responsibility, which rated equal third with "work culture" sub-theme.

In the following Chap. 8, "Discussion", a detailed appraisal has been carried out of these four major sub-themes, from which conclusions have been drawn and recommendations made, for this research surrounding sources of site waste.

7.3 Front-End Site Waste Minimization Strategy Findings

Section 7.3 considers survey questionnaire Part 3, Questions 26–34, inclusive, in which the 48 respondents were asked to comment on possible waste minimization strategies, using an abridged version of the Yates (2013) survey questionnaire on waste minimization, to determine whether these initiatives are being adopted on Australian projects. The focus of this adapted questionnaire was on upstream site reduction strategies, rather than downstream construction materials supply chain considerations, such as recycling of waste materials and/or transport of residual waste to landfill. The questionnaire template, having been based on Yates' (2013) survey, was suitably abridged to include only those questions related to front-end waste reduction

If respondents answered in the affirmative to using one of the strategies nominated in Questions 26–33, inclusive, then further comments were solicited. Question 34 requested respondents to provide information on waste minimization strategies adopted on their projects other than those nominated in Questions 26–33, inclusive. The 48 respondents to Part 3 survey answered Questions 26–34, inclusive, of the questionnaire covering site waste minimization. The Yates (2013) paper diligently recorded pertinent respondent comments in their paper and noted that the entire suite of participant remarks was included in their research project data acquisition knowledge base on site waste minimization practices internationally. Yates (2013) conducted their study via literature review, and more particularly in the USA, via a survey of engineering construction industry top management cohort.

As an adjunct to enhance this research, a qualitative thematic analysis of this data was undertaken as the best qualitative methods approach, to examine respondent comments for each of the Questions 26–34, inclusive, in Part 3 survey on site waste minimization.

7.3.1 Thematic Analysis Findings Comprising Coding/Themes/Sub-themes for Part 3 Front-End Waste Minimization Strategies

Tables 7.14, 7.15 and 7.16 in this section describe and summarize the thematic analysis used in this qualitative research, which has focused on evaluating the themes captured within the data. See Sect. 7.2.4, "Thematic Analysis of Part 2 Sources of Site Waste Using Coding/Themes/Sub-Themes", for a detailed description of the thematic analysis approach to qualitative research including coding, themes and sub-themes.

Question 26—Interpretation of Key Findings
Question 26 asked respondents: "*is your firm using techniques that improve resource efficiency, equipment efficiency and material resource efficiency, and which allow for training of manual labour?*"
    To which the 48 respondents answered:

  I. "No" for 39.58 per cent;
 II. "Don't know" for 18.75 per cent; and
III. "Yes" for 41.67 per cent.

Table 7.14 Question 26, Part 3 waste minimization thematic analysis

No.	Respondent comments—Question 26, Part 3 waste minimization	Code	Theme	Sub-theme
1	Machine systems control	N/A	N/A	
2	Planning and scheduling latest techniques utilized, e.g. use barges for spoil haulage, to minimize truck movements in the CBD	Innovative engineering construction processes	Management	–
3	We are (our company) more and more improving design detailing from project lessons learned, to make sure re-work is minimized and waste from error is reduced	Good early design	Design	FEED
4	Digital BIM engineering	Innovative engineering construction processes	Management	–
5	Multiple skips for sources of separation	Adequate recycling resources	Management	Recycling resources
6	Very early involvement in the design and off-site fabrication	Early design involvement	Design	FEED
7	Reviewing site collaboration software to enhance communication and timeframes for accessing information	Innovative communication process	Management	Innovative project management ICT support
8	Rigorous inspections at all stages; realistic material delivery schedules	Supply chain	Procurement	Quality process shortfall

(continued)

Table 7.14 (continued)

No.	Respondent comments—Question 26, Part 3 waste minimization	Code	Theme	Sub-theme
9	Apps on mobile devices to replace paperwork for surveillance; assessment; HSE	Paperless site communication	Management	Innovative project management ICT support
10	Using proper estimating system (like D'Est software) provides immediate insight into material requirements, through the procurement sector of the programme	Innovative estimating systems that can provide real-time material estimates	Management	Innovative project management ICT support
11	Better quality trade and supervision reduce mistakes and waste	Better tradespeople/better supervision	Workers/management	Work culture/training
12	Better supply chain management	Supply chain	Procurement/management	Integrated supply chain/innovative project management ICT support
13	Off-site modularization	Good early design/select less traditional stick build	Design	FEED
14	The company I work for uses excellent waste strategies in waste reduction, most times. Recycling bins of exact metal types and the consolidation of (scrap) large nuts, bolts, studs, etc.	Adequate recycling resources	Recycling	Recycling resources

(continued)

Table 7.14 (continued)

No.	Respondent comments—Question 26, Part 3 waste minimization	Code	Theme	Sub-theme
15	We are designers and site supervisors and do not have a direct effect on construction materials procurement, delivery and utilization/construction, to manage or affect contractor waste	Consultant abrogates responsibility to contractor per waste	Management	Work culture/ responsibil-ity/training
16	Integrated supply chain; "just-in-time" deliveries	Supply chain	Procurement	Integrated supply chain
17	Equipment selection; training; process planning and incentives	Positive behaviour enhancement for entire design to build project team	Management	Work culture/training
18	Recommending to clients to include a procurement baseline from concept through to commissioning that includes sustainable use of materials, in both construction and ongoing maintainability	Supply chain	Procurement	Integrated supply chain/client responsibility

By P. Rundle, February 2018

Table 7.15 Themes for waste minimization Part 3, Question 26

Theme No.	Theme description	Number of times this theme was recorded against respondent comment
1	Management	10
2	Procurement	4
3	Design	3
4	Workers	1
5	Recycling	1

By P. Rundle, February 2018

Table 7.16 Sub-themes for waste minimization Part 3, Question 26

Sub-theme No.	Sub-theme description	Number of times this sub-theme was recorded against respondent comment
1	Innovative project management ICT support	4
2	FEED	3
3	Integrated supply chain	3
4	Work culture	3
5	Training	3
6	Client responsibility	2
7	Recycling resources	2
8	Quality process shortfall	1

By P. Rundle, February 2018

Section 1.2 of this textbook specifically defines "efficiency" in the context of this research, as: "meeting all internal requirements for cost, margins, asset utilization and other related efficiency measures" (Sundqvist et al. 2014). The respondent survey questionnaires provide a frank evaluation of their firm's drive for competitive advantage. Only 41.67 per cent of participants indicated that, "yes", their company was striving to be more efficient. An almost equal number of respondents answered "no", with 39.58 per cent of employees simply not knowing whether their company is striving to be more efficient with resources, personnel, materials and equipment. Table 6.4 respondent answers to Part 3 survey, Questions 26–34, notes that on average, 46.07 per cent of respondents answered "yes" on all survey questions regarding waste minimization strategies. This validates the author's findings on a relatively low level of respondent confidence that Australian engineering construction, in the participant's sphere at least, is not being undertaken particularly "efficiently".

Table 7.14, Question 26, "Themes", shows that the "management" theme is perceived as the most important factor that impacts on efficiency. This is a mature assessment by the 48 respondents because many, if not most, respondents are from the management cadre; indeed, there are 16 company directors, including several chief executive officers and one chair, who participated in this research survey. Often, poor worker productivity is blamed solely on the performance of the workers (Chignell 2017, n.p.). However, a root cause analysis of a firm, a project, or a work groups' productivity can most frequently be traced back to poor management practices (Heikkinen 2017).

Table 7.15, Question 26, "Sub-themes", indicates a very flat spread across several sub-themes. However, the author's interpretation of the thematic analysis results is that there are three key drivers to improving efficiencies in the utilization of physical and non-physical resources, as follows:

I.   Out of all nine questions in Part 3 of this waste minimization questionnaire, Question 26 is the only time that the subject of "innovative project management ICT support" was seen as the single major impacting sub theme by survey participants. These respondents acknowledge that keeping up to date with project management ICT is the key to efficient use of material, equipment and personnel resources, which will reduce site waste (Cavalante 2013).
II. There is considerable technical and logistical expertise required to, respectively, establish robust "front-end engineering design" processes (Rockwell 2015) and an efficient/effective "integrated supply chain" (Jiamei 2012). Therefore, these are two mechanistic areas that reside within the capabilities of competent contractors/builders.
III. Survey participants perceptively recognized that both "work culture" and "training" go hand in hand; improving management, worker efficiencies and site waste reduction will be a vital adjunct to the need to relentless improving efficiencies to staying in a competitive business (Igo and Skitmore 2006).

7.3.1.1 Codes/Themes/Sub-themes for Respondent Comments for "Yes" Answer to Question 27, Part 3 Waste Minimization

See Tables 7.17, 7.18 and 7.19.

Question 27—Interpretation of Key Findings

Question 27 asked respondents: *"are innovative designs, construction components or construction processes being integrated into your projects to reduced site generated waste?"*, to which the 48 respondents answered:

I.   "No" for 31.25 per cent;
II. "Don't know" for 20.83 per cent; and
III. "Yes" for 47.92 per cent.

Clearly just under half of the survey respondents opined that innovative designs and construction methodologies were being employed in their particular Australian engineering construction area.

It is worth recapping the literature review on engineering construction innovation, found in Sect. 2.1, "Option Selection Background", of this textbook, which indicates that there has been a degree of academic innovation in the field of construction engineering, particularly in OECD countries (Suprun and Stewart 2015). Innovation usually leads to reduced construction costs and/or project schedule improvements (NRUHSE 2013). Gann (2000) suggests that higher levels of construction innovation shall improve productivity, whilst Tatum (1991) posits that profitability will increase as a result of employing innovative construction techniques. The international construction sector has been defined as a "laggard industry" (Dibner and Lerner 1992)

Table 7.17   Question 27, Part 3 waste minimization thematic analysis

No.	Respondent comments to Question 27—Part 3 waste minimization	Coding	Theme	Sub-theme
1	Embedding in the Design Management Plan	Design for waste mitigation	Design	FEED
2	Promoting single package bundling of multiple components rather than multiple, individual wrapped components. Delivery and interim storage in containers to provide protection until use—results in reduction of weatherproof packaging	Reduce excess packaging	Procurement	Innovative packaging solutions
3	Off-site prefabrication in factory environment where possible	Maximize off-site prefabrication	Management	Maximize off-site construction fabrication and pre-assembly
4	Modular construction strategy	Off-site modularization and pre-assembly	Management	Maximize off-site construction fabrication and pre-assembly
5	Yes, but not directly to reduce waste but to provide an efficient design. It is assumed that the contractor will manage waste as this would reflect in construction cost and time reductions	Design for waste mitigation	Design	Client responsibility (in this case to distribute risk)
6	Limited scale depending on contract (e.g. are alternative design/solutions accepted) and alternative components approval processes	Design for waste mitigation	Design	Client responsibility (to distribute risk)
7	Off-site fabrication. Quality detailed design. Good supervision. Planning	Design for waste mitigation/maximize off-site prefabrication/supervision	Management/design	Maximize off-site construction fabrication and pre-assembly
8	Lost formwork, integrated structural elements, bond deck	Prudent formwork selection	Management	Innovative formwork solutions
9	As per Question 26 previously (cannot match anonymous response)	–	–	–

(continued)

Table 7.17 (continued)

No.	Respondent comments to Question 27—Part 3 waste minimization	Coding	Theme	Sub-theme
10	Being aware of potential waste generation and taking the time to minimize it	Training on waste source identification to minimize	Management	Training
12	Modularization and maximized pre-assembly and precast are the keys to reducing construction waste	Off-site modularization and pre-assembly/precast concrete	Management	Maximize off-site construction fabrication and pre-assembly
13	Approval documents are being developed to focus on enabling companies to maximize opportunities to use the most efficient designs and construction methodologies by focusing compliance requirements on the outcome to be achieved rather than the process to achieve compliance	Seek client advice on preliminary designs and build methods	Management	FEED
14	Typically driven by constructor initiatives prior to design finalization. Significant room for improvement by design consultants	Seek client advice on prelim. designs and build methods/good front-end design	Management/design	FEED
15	Utilization of Davit and tower cranes	Optimize crane selection	Management	–
16	We are constantly looking for innovative design approaches to give us an advantage and that includes minimizing wasted time	Early contractor involvement	Management	Innovative technical solutions
17	Greater use of prefabrication off-site	Off-site prefabrication	Management	Maximize off-site construction fabrication and pre-assembly
18	Very early involvement in the design, input in the design company experts, implementation of off-site fabrication, partial off-site pre-assembly of building elements	Early contractor involvement/off-site prefabrication/pre-assembly	Management	Maximize off-site construction fabrication and pre-assembly

(continued)

Table 7.17 (continued)

No.	Respondent comments to Question 27—Part 3 waste minimization	Coding	Theme	Sub-theme
19	Ground stabilization product that enable use on "unsuitable" engineering soil as useable material	Use unique technology to make use of unsuitable materials and incorporate into the works	Management	Innovative technical solutions
20	As per above, innovative design solutions for us means "smarter and easier" construction in the field and this reduces waste through error in a tough and competitive commercial building sector in Melbourne	Innovative design as matter of course	Design	Innovative technical solutions
21	3D design tools	Innovative design tools	Management	Innovative technical solutions
22	Seeking resource recovery exemptions to maximize the amount of materials that can be re-used; recycling targets set in contract requiring high percentage of construction and demolition waste to be recycled, and spoil to be re-used; negotiate take-back agreements with suppliers for packaging; maximize use of recycled or re-used materials (e.g. wood, steel, site facilities)	Innovative recycling/reduce packaging	Recycling/procurement	Adopt innovative commercial arrangements/innovative packaging solutions
23	We have a waste minimization policy which is in development (N/A comments)	–	–	–

By P. Rundle, February 2018

Table 7.18   Themes for waste minimization Part 3, Question 27

Theme No.	Theme description	Number of times this theme was recorded against respondent comment
1	Management	15
2	Design	4
3	Procurement	2
4	Recycling	1

By P. Rundle, February 2018

Table 7.19   Sub-themes for waste minimization Part 3, Question 27

Sub-theme No.	Sub-theme description	Number of times this sub-theme was recorded against respondent comment	Remarks
1	Maximize off-site construction fabrication and pre-assembly	6	
2	Innovative technical solutions	5	
3	FEED	3	Two comments stressed need for early contractor involvement (ECI) at preliminary design phase
4	Innovative packaging solutions	2	There are two excellent and original concepts from two respondents on this sub-theme with packaging identified under sources of site waste as a major cause. Both concepts shall be incorporated into recommendations
5	Client responsibility	2	Both comments refer to apportionment of risk which is ultimately the client's responsibility to apportion and would be best handled collaboratively with contractor/designer
6	Innovative formwork solutions	1	
7	Training	1	

By P. Rundle, February 2018

Table 7.20 Part 3 waste minimization thematic analysis

No.	Respondent comments to Question 28, Part 3 waste minimization	Coding	Theme	Sub-theme
1	Design and engineering use the latest information in the industry	Design and engineering use latest technology	Management	Innovative technological solutions
2	We have waste management plans, but their scope is limited to contractor activities	Only contractor has waste management plan responsibility	Management	Client responsibility
3	Early contractor involvement (ECI)	–	Management	FEED
4	Shipping container contents are considered during engineering	Design to account for shipping parameters at early design stage	Design	FEED
5	Attention to detail in design and shop detailing	Detailed design	Design	–
6	Structurally efficient	Detailed design maximizes structural steel	Design	–
7	We pay considerable attention to early design in the definitive stage	Definitive feasibility study picks up early design	Design	FEED
8	Value engineering in design	Value engineering at early design phase	Design	FEED
9	For design and build contracts where there is economic incentive to design out waste or in response to sustainability rating schemes (green building/infrastructure)	Initiate value engineering and savings sharing incentive	Design	FEED/client responsibility
10	Constructability reviews	Review base designs for ease of construction	Design	FEED

(continued)

Table 7.20 (continued)

No.	Respondent comments to Question 28, Part 3 waste minimization	Coding	Theme	Sub-theme
12	Part of method statement development	Review construction methodology on base designs	Design	FEED
13	This emanates from PM, QMS, staging and construction planning	Early design-phase project plan review	Design	FEED
14	Is taken into consideration (N/A)	–	–	–
15	A project-specific planning assessment as to programme preparation and product availability	Early design-phase project plan review	Design	FEED
16	Thoughtful design development and documentation	Early design and engineering considerations	Design	FEED
17	We have a structured approach—it is immature, and we are starting to embrace LEAN technologies to improve our approach	Adopt lean construction engineering approach which highlights integrated supply chain production line approach	Management/ procurement	Innovative technol-gical solutions/integrate - upply chain
18	Planning of tunnel spoil disposal in landfill and balancing cut and fill in road construction	Develop prudent design/build methodology at early planning phase	Design	FEED
19	Maximize employment of full pre-assembly	Pre-assembly	Management	Maximize off-site construction fabrication and pre-assembly
20	Regular design and implementation reviews. Running of a risk and opportunity register	Capture design and build reviews from inception and use risk-based decision-making	Design	FEED
21	Embedded in the Design Management Plan	Waste reduction part of design guidelines	Design	FEED

By P. Rundle, February 2018

Table 7.21 Themes for waste minimization Part 3, Question 28

Theme No.	Theme description	Number of times this theme was recorded against respondent comment
1	Design	14
2	Management	5
3	Procurement	1

By P. Rundle February 2018

Table 7.22 Sub-themes for waste minimization Part 3, Question 28

Sub-theme No.	Sub-theme description	Number of times this sub-theme was recorded against respondent comment	Remarks
1	FEED	13	One comment stressed the need for ECI in early design phase Another comment highlighted the need for a formal risk management process to be adopted in the design process
2	Innovative technical solutions	2	
3	Client responsibility	2	
4	Maximize off-site construction fabrication and pre-assembly	1	
5	Integrated supply chain	1	

By P. Rundle February 2018

in terms of innovative development. This is because of the inherent high-risk nature, along with the low margin nature, of the work (Slaughter 1993). Several studies have denoted the importance of innovation in improving construction outcomes and yet have highlighted the apparent disconnect with the failure to diffuse these new innovations into the construction industry (Davidson 2013; Egbu 2004; Sayfullina 2010).

National spending on innovation is not great, even in OECD countries. As an example, Australia spends only 2.4 per cent of its gross domestic product (GDP) on research and development (R&D), whilst the USA spends 2.9 per cent on its R&D (Hampson et al. 2014). Coupled with the inherent risk adverse nature of the construction business, firms are conservative in their approach to innovative solutions (Miller et al. 2009). Accordingly, the diffusion of all available knowledge on efficient and effective construction methodologies available to the construction world, rather than an explosive boost in construction innovation research and development via a dramatic increase in expenditure, would prove an altogether more practical means of addressing this situation (Slaughter 1993).

During discussions regarding innovation potential of the four "short-list" options, in the "Go Forward" case peer review meeting of the 17 July 2018, panellist, G. Stacey, suggested that innovation on engineering construction projects is most usually driven by a chaotic event such as a construction collapse in the first instance, followed by innovative material/equipment vendor research, and lastly via contractor/designer innovations (Peer Review Meeting Two 2018).

Unambiguously, Table 7.18 theme summation shows that the "management" theme is viewed by participants as the predominant influencer on innovation, with 15 respondent comments, followed by six comments from respondents highlighting the importance of the "design" theme. Table 7.19 sub-theme summation shows that there were six respondent comments on the "maximize construction fabrication and pre-assembly" sub-theme; with five respondent comments remarking on the need for the provision of the "innovative technical solutions" sub-theme and the third sub-theme indicating a requirement for the "front-end engineering design" sub-theme. These sub-themes covering, (i) "maximize construction fabrication and pre-assembly"; (ii) "innovative technical solutions"; and (iii) "front-end engineering design", are consistently interwoven threads within the participant narratives collated from the survey questionnaire responses as important methods of reducing construction site waste.

The findings in this sub-section suggest that further academic research will be necessary to determine if there is an adequate knowledge transfer of academic research into innovative engineering construction from academia to industry. Furthermore, researching exactly what division of innovation is between disruptive, vendor and planned, and to this end, an abridged version of the seminal Slaughter model (1993), could be used (see Sect. 2.1, "Option Selection Background") to validate these questionnaires against historical data. This future research data will add to the knowledge acquisition database on site waste minimization to further develop appropriate project waste strategies through innovative off-site modularization. Early front-end design to allow for waste minimization, via precast concrete for example, and innovative technical solutions are seen as being key initiatives in site waste minimization, with management and design being the major drivers of innovation.

7.3.1.2   Codes/Themes/Sub-themes for Respondent Comments for "Yes" Answer to Question 28, Part 3 Waste Minimization

Question 28—Interpretation of Key Findings
See Tables 7.20, 7.21 and 7.22.

Question 28 asked respondents: "*do you adopt a structured approach both for engineering design and in the determination of construction methodologies that involves waste minimization strategies?*", to which the 48 respondents answered:

Table 7.23 Question 29, Part 3 waste minimization thematic analysis

No.	Respondent comments to Question 29, Part 3 waste minimization	Coding	Theme	Sub-theme
1	Yes. Precast is used where appropriate and steel table formwork/falsework for high-rise soffits	Precast/prudent formwork solutions	Management	Maximize off-site construction fabrication and pre-assembly/innovative formwork solutions
2	Yes. Precasting concrete units and fabrication of large electrical and mechanical units/products in factories/manufacturing facilities	Optimized precasting and off-site assemblies	Management	Maximize off-site construction fabrication and pre-assembly
3	We definitely are being asked to consider more modularized options for our engineering and architectural designs in the industrial sectors for manufacturing and mining	Early attention at design-phase to off-site fabrication being requested by clients	Management	FEED/maximize off-site construction fabrication and pre-assembly/client responsibility
4	Used intermediate bulk containers are donated to people requiring water storage off-grid	Practical recycle of non-permanent materials for community benefit	Recycling	Recycling resources
5	Use prefabrication where possible	Maximize prefabrication	Management	Maximize off-site construction fabrication and pre-assembly
6	Strong drive for prefabrication, pre-assembly and modularization, e.g. prefabrication HV sub-stations delivered to site	Prefabrication such as sub-stations is being pushed by clients	Management	Maximize off-site construction fabrication and pre-assembly/client responsibility
7	Strong focus on modularization and off-site pre-assembly	Maximize prefabrication	Management	Maximize off-site construction fabrication and pre-assembly
8	Prefabrication and pre-assembly—up to the Subbie to manage waste	Sub-contractor's responsibility for prefabrication	Management	Maximize off-site construction fabrication and pre-assembly/client responsibility
9	Prefabrication is maximized to the limits of transport	Optimized use of prefabrication	Management	Maximize off-site construction fabrication and pre-assembly

(continued)

Table 7.23 (continued)

No.	Respondent comments to Question 29, Part 3 waste minimization	Coding	Theme	Sub-theme
10	Prefabrication and modularization used; selection is project-dependent and varies between projects	Optimized use of prefabrication	Management	Maximize off-site construction fabrication and pre-assembly
12	Precast, prefabrication and modularization used	Optimized use of prefabrication, precast and modules	Management	Maximize off-site construction fabrication and pre-assembly
13	Precast/modular construction techniques; site measurement	Optimized use of prefabrication, precast and modules	Management	Maximize off-site construction fabrication and pre-assembly
14	Precast or modularization is commonly used	Optimized use of prefabrication, precast and modules	Management	Maximize off-site construction fabrication and pre-assembly
15	Off-site prefabrication and modularization	Optimized use of prefabrication and modules	Management	Maximize off-site construction fabrication and pre-assembly
16	Off-site construction is more efficient than on-site	Optimized use of prefabrication, precast and modules	Management	Maximize off-site construction fabrication and pre-assembly
17	Most projects push the responsibility of waste management to EPC or EPCM who in turn push it further down the commercial pecking order and finally to landfill	Limits of responsibility for site waste delegated down	Management	Client responsibility (to determine risk)
18	Modularization and bulk packaging	Optimize for module off-site construction/work with suppliers to provide bulk packaging	Management/ procurement	Maximize off-site construction fabrication and pre-assembly/innovative packaging solutions
19	Identification of suitable spoil disposal areas	Early identification of pits, site won materials and spoil dumps critical to design effort	Design	FEED
20	Current project utilizing precast components, including precast factory on site	Optimize precast	Management	Maximize off-site construction fabrication and pre-assembly

(continued)

Table 7.23 (continued)

No.	Respondent comments to Question 29, Part 3 waste minimization	Coding	Theme	Sub-theme
21	Construction methodology well done by experienced people	Early evaluation of build methods	Design	FEED
22	Constantly looking to minimize waste—precast, retain existing infrastructure, avoid temporary works and side-tracks, etc.	Optimize precast	Design	Maximize off-site construction fabrication and pre-assembly
23	Assessing the design to suit logistics (roads transport/lifting capacities), implementation of off-site manufacturing, off-site partial pre-assembly	Early design to aid logistics, including pre-assembly access	Design	FEED

By P. Rundle, February 2018

Theme No.	Theme description	Number of times this theme was recorded against respondent comment
1	Management	17
2	Design	4
3	Procurement	1
4	Recycling	1

Table 7.24 Themes for waste minimization Part 3, Question 29

By P. Rundle, February 2018

I. "No" for 29.17 per cent;
II. "Don't know" for 20.83 per cent; and
III. "Yes" for 50.00 per cent.

In accordance with the above findings, half of the respondents were of the opinion that their respective company's approach to waste minimization was in a structured manner, primarily in the designs area (14 design theme comments), with five respondents highlighting the "management" theme as being the next driver. Overwhelmingly, half of the "yes" respondents believed that "FEED" was the major sub-theme covering structured approaches to waste minimization.

7.3.1.3 Codes/Themes/Sub-themes for Respondent Comments for "Yes" Answer to Question 29, Part 3 Waste Minimization

Question 29—Interpretation of Key Findings
See Table 7.23.

Question 29 asked respondents: *"do you address waste generation reduction during project pre-planning to utilize designs that minimize waste using any of the following techniques: precast; prefabrication; pre-assembly or modularization?"*, to which the 48 respondents answered:

I. "No" for 29.17 per cent;
II. "Don't know" for 14.58 per cent; and
III. "Yes" for 56.25 per cent.

The survey results from the 48 respondents showed that the majority of participants (56.25 per cent) indicated that their companies adopted waste reduction processes at the pre-planning stage by considering such methodologies as precast concrete elements, modular construction and off-site assembly. Table 7.24 shows that 17 respondents believed that the "management" theme was the key driver in pre-planning for waste minimization. Table 7.25 notes that the "maximize off-site construction fabrication and pre-assembly" sub-theme is the major pre-planning factor when considering FEED requirements for waste minimization.

Table 7.25 Sub-themes for waste minimization Part 3, Question 29

Sub-theme No.	Sub-theme description	Number of times this sub-theme was recorded against respondent comment	Remarks
1	Maximize off-site construction fabrication and pre-assembly	17	
2	FEED	4	
3	Innovative packaging solutions	1	Comment constructively looks at issue and shall be included in recommendations
4	Innovative formwork solutions	1	
5	Client responsibility	1	Comment is a project risk apportionment determination matter

By P. Rundle February 2018

7.3.1.4 Codes/Themes/Sub-themes for Respondent Comments for "Yes" Answer to Question 30, Part 3 Waste Minimization

Question 30—Interpretation of Key Findings

Question 30 asked respondents: "*as builders, contractors and engineering consultants, do you ensure a minimum amount of permanent and temporary materials are expended in the effective provision of client conforming construction/building products?*"

To which the 48 respondents answered (Table 7.26):

 I. "No" for 31.25 per cent;
 II. "Don't know" for 31.25 per cent; and
III. "Yes" for 37.50 per cent.

This question had the lowest respondent "yes" rate of 37.5 per cent of respondents. This is equal to the number of respondents who replied in the affirmative to Question 33, regarding whether their company had a waste minimization plan.

There are some comments in the respondent answers to other questions in this survey, in all probability from consultants and/or clients, which are critical of contractors passing waste material costs directly onto the clients, often in the form of variations.

Table 7.27 shows that the predominant driver for this question is the "management" theme, whilst Table 7.28 indicates that five participants responded in such a way that inferred that the "contractor presumption based on client pays" was the

Table 7.26 Question 30, Part 3 waste minimization thematic analysis

No.	Respondent comments to Question 30, Part 3 Waste minimization	Coding	Theme	Sub-theme
1	Minimizing materials minimizes cost	Contractor posits it is optimizing materials to maintain competitive advantage	Management	Contractor presumption based on client pays?
2	Waste minimization and sustainable materials strategies implemented on project	Contractor posits it is optimizing materials to maintain competitive advantage	Management	Contractor presumption based on client pays?
3	Part of our sustainability model	Allows in design for waste minimization	Design	FEED
4	Minimize overorder and retain excess for inclusion in the next job	Minimize overordering	Procurement	Quality process shortfall
5	Sometimes specify recycled materials, e.g. glass sand	Specify recycled product	Procurement	Client responsibility
6	Methodology of erection/temporary works is designed at the same time as permanent elements of the building	Early design of form-work/falsework	Design	FEED
7	If we do not do this, then we do not win work	Contractor posits it is optimizing materials to maintain competitive advantage	Management	Contractor presumption based on client pays?
8	Not sure as to what extent (N/A response)	–	–	–
9	Waste minimization is part of conscientious cost control	Contractor posits it is optimizing materials to maintain competitive advantage	Management	Contractor presumption based on client pays?
10	By conducting value engineering exercises	Value engineering at early design phase	Design	FEED

(continued)

Table 7.26 (continued)

No.	Respondent comments to Question 30, Part 3 waste minimization	Coding	Theme	Sub-theme
12	Usually anything temporary is found a new "home/purpose" somewhere else in the project	Recycle materials within site for internal re-use	Recycling	Recycling resources
13	As per Question 28 above (anonymous response so N/A)	–	–	–
14	Prefabrication and quality installation contracts	Prefabrication optimizes materials	Management	Maximize off-site construction fabrication and pre-assembly
15	Actions are taken to minimize material use; often, this is outside of our control and is the building contractor. This often results in designs presented for approval, which could be modified to reduce waste, but time constraints prevent the additional cycle to modify the design	Need for all front-end design reviews for waste reduction with client	Management	FEED/client responsibility
16	Where there is opportunity through the contract and mutual benefit (e.g. green building)	If the contractor is paid, then they will optimize their waste savings	Management	Client responsibility
17	We deliver to the client a healthy standard of workmanship and minimal wastage to minimize actual costs	Contractor posits it is optimizing materials to maintain competitive advantage whilst providing conforming product	Management	Contractor presumption based on client pays?

By P. Rundle, February 2018

Table 7.27 Themes for waste minimization Part 3, Question 30

Theme No.	Theme description	Number of times this theme was recorded against respondent comment
1	Management	8
2	Design	3
3	Procurement	2
4	Recycling resources	1

By P. Rundle, February 2018

Table 7.28 Sub-themes for waste minimization Part 3, Question 30

Sub-theme No.	Sub-theme description	Number of times this sub-theme was recorded against respondent comment
1	Contractor presumption based on client pays?	5
2	FEED	4
3	Client responsibility	3
4	Quality process shortfall	1
5	Maximize off-site construction fabrication and pre-assembly	1
6	Recycling resources	1

By P. Rundle, February 2018

top sub-theme selection. The "FEED" sub-theme (four responses) and the "client responsibility" sub-theme (three responses) closely followed the aforementioned sub-theme.

7.3.1.5 Codes/Themes/Sub-themes for Respondent Comments for "Yes" Answer to Question 31, Part 3 Waste Minimization

Question 31—Interpretation of Key Findings

Question 31 asked respondents: *"regarding the use of temporary construction materials, do you consider waste minimization processes? For example, for concrete construction, do designers specify concrete elements of similar dimensions, where practical; are steel shutters used on repetitive formwork; is formwork adequately*

Table 7.29 Question 31, Part 3 waste minimization thematic analysis

No.	Respondent comments to Question 31, Part 3 waste minimization	Coding	Theme	Sub-theme
1	To an extent, it is risky to run full JIT	"Just-in-time" ordering has potential schedule risk	Procurement	Innovative project management ICT support
2	Maximize use of recycled or re-used materials (e.g. wood, steel, site facilities)	Maximize re-use of temporary works and materials	Recycling	Recycling resources
3	We use most cost-effective concrete production techniques that also consider worker knowledge of method; steel shutters, etc. Quick construction, with no re-work maintains profits, finishes projects on time and reduces waste	Prudent formwork selection using worker knowledge	Management/workers	Innovative formwork solutions/training
4	Re-use formwork where possible	Robustly construct and importantly maintain formwork	Workers	Training
5	Have previously developed reusable forms for spread footings and formwork	Prudent formwork selection	Management	Innovative formwork solutions
6	Best use of available formwork systems	Prudent formwork selection	Management	Innovative formwork solutions
7	Very tight and accurate programming	Accurate planning and scheduling reduces waste	Management	Innovative project management ICT support
8	Always looking to simplify and avoid temporary works costs that do not value add	Prudent formwork and falsework selection	Management	Innovative formwork solutions

(continued)

Table 7.29 (continued)

No.	Respondent comments to Question 31, Part 3 waste minimization	Coding	Theme	Sub-theme
9	This is standard practice	Prudent formwork selection and off-site fabrication are industry standard	Management	Innovative formwork solutions/training
10	By specifying the type of formwork to be used	Prudent formwork selection	Management	Innovative formwork solutions
12	Standardization is virtually non-existent. Precast is almost an evil conjuring. How could the large engineering houses make profit from cookie cutter approach. The baulk at pre-assembly because it minimizes post-design-phase follow-up works	Engineering houses discourage standard designs and prefabrication/precast techniques as it reduces design hours	Management/ design	Maximize off-site construction fabrication and pre-assembly/FEED
13	Dictated by the MC or contractor generally	Managing contractor only dictates optimal methodologies	Management	Client responsibility
14	Prefabrication wherever possible. Delivery "just in time". Construction planning.	Use a combination of available techniques as per prefabrication, JIT and planning	Management/ procurement	Maximize off-site construction fabrication and pre-assembly/integrated supply chain/innovative project management ICT support
15	As a PMC, these are not directly controlled or specified in contracts. However, selected sub-contractors (e.g. concrete construction) do make use of re-usable formwork	Sub-contractor dictates formwork use not managing contractor	Management	Client responsibility
16	Standardization of design elements, re-use of temporary construction materials	Standardize structural elements and re-use temporary materials, efficiently	Design/ recycling	FEED/recycling resources

(continued)

Table 7.29 (continued)

No.	Respondent comments to Question 31, Part 3 waste minimization	Coding	Theme	Sub-theme
17	Where feasible, often clients design may not suit that approach (non-standard elements, etc.)	Standardization often not possible due to client "one-off" design	Management/design	Client responsibility/FEED
18	Yes and no depending on the job contractor and or client in certain situations yes in others quite the opposite	Standardization often not possible due to client "one-off" design	Management/design	Client responsibility/FEED
19	Yes—repetitiveness but for cost efficiency resulting in re-use of temporary works	Standardize structural elements and re-use temporary materials, efficiently	Management/design	Client responsibility/FEED
20	"Just-in-time" orders, standardization of concrete structures	Use a combination of available techniques as per standardization and JIT	Design/procurement	FEED/integrated supply chain
21	Permanent formwork; steel and aluminium scaffold re-use	Optimize formwork/falsework	Management	Innovative formwork solutions
22	Re-use of formwork and falsework critical to prudent temporary works design and installation for commercial reasons	Optimize formwork/falsework	Management	Innovative formwork solutions

By P. Rundle, February 2018

Table 7.30 Themes for waste minimization Part 3, Question 31

Theme No.	Theme description	Number of times this theme was recorded against respondent comment
1	Management	16
2	Design	6
3	Procurement	3
4	Workers	2
5	Recycling	2

By P. Rundle, February 2018

treated and robustly fabricated, to allow re-use and are orders "just in time" to reduce material losses on site?", to which the 48 respondents answered (Table 7.29):

  I. "No" for 18.75 per cent;
 II. "Don't know" for 29.17 per cent; and
III. "Yes" for 52.08 per cent.

Regarding waste minimization of temporary construction materials, 52.08 per cent of respondents indicated that their company used appropriate practices to reduce waste from these materials such as formwork/falsework; however, 18.75 per cent indicated that their company did not use appropriate practices. Table 7.30 summarizes themes, of which the major driver was the "management" theme with 16 respondent comments, followed by the "design" theme (six comments) and the "procurement" theme (three comments).

The majority of Part 3 questions on waste minimization thematic analysis have shown that, by and large, in descending order, the top theme has been "management", followed by "design" and finally the "procurement" theme. Table 7.31, summarizing sub-themes for this Question 31, notes that the "innovative formwork design" sub-theme is the main factor for this question, with eight respondent comments. The other two significant sub-themes are the "FEED" sub-theme (six comments) and the "client responsibility" sub-theme (five comments). The fourth sub-theme with three respondent comments is the "innovative project management ICT support" sub-theme, which all participants felt was necessary to ensure that the utilization of JIT procurement of materials would need to be carefully controlled to avoid procurement delays.

7.3.1.6 Codes/Themes/Sub-themes for Respondent Comments for "Yes" Answer to Question 32, Part 3 Waste Minimization

Question 32—Interpretation of Key Findings

Question 32 asked respondents: *"do the contractors/builders, consultants and ven-*

Table 7.31 Sub-themes for waste minimization Part 3, Question 31

Sub-theme No.	Sub-theme description	Number of times this sub-theme was recorded against respondent comment	Remarks
1	Innovative formwork solutions	8	
2	FEED	6	
3	Client responsibility	5	
4	Innovative project management ICT support	3	Three comments alluded to "just-in-time" material deliveries and one of these comments made the point about how the risk potential for this process needed to be managed. Integrated ICT will mitigate this risk
5	Maximize off-site construction fabrication and pre-assembly	2	
6	Integrated supply chain	2	
7	Training	2	
8	Recycling resources	2	

By P. Rundle, February 2018

dors, constructively work with the client to minimize change orders that make preordered products redundant and suitable only for waste?", to which the 48 respondents answered (Table 7.32):

   I. "No" for 35.41 per cent;
  II. "Don't know" for 22.92 per cent; and
 III. "Yes" for 41.67 per cent.

This is the second lowest scored question in the "yes" category for all Part 3 waste minimization questions, at a 41.67 per cent affirmative response rate, with 16 respondents signifying that "management" was the primary driver to constructively address this waste reduction strategy. This is of concern because in Part 2 of this survey on sources of waste, it was found that the most consistent highest scoring source of site waste was "client-initiated design changes" (i.e. variations). The next highest driver was the "design" theme, with six comments, and then "procurement" (three comments). Table 7.33 shows that the primary theme for this question is "management", with 13 respondent comments, followed by "procurement" with two comments. Sub-theme factors, detailed in Table 7.34, clearly indicate that the

Table 7.32 Question 32, Part 3 waste minimization thematic analysis

No.	Respondent comments to Question 32, Part 3 waste minimization	Coding	Theme	Sub-theme
1	Many coordination meetings and training	Coordination meetings with client and training, minimize change orders	Management	Communication/training
2	N/A to current project	Client interface not part of current project	Management	Communication/client responsibility
3	Through collaboration "from day one"	Collaborate with client all through project	Management	Communication
4	Where possible, we will aim to have weekly interaction meetings to keep things on track and focused	Weekly meeting with client/contractor	Management	Communication
5	"Managing the client" is still a challenge	Managing client relations requires skills	Management	Communication/training
6	If obliged to do so in accordance with the consultant's technical specification	Interface with client only if the consultants' brief demands	Management	Client responsibility/communication/training
7	Any variation has to be carried out to lowest cost. The contractor does not overorder, as the client usually will not pay reasonable costs in every case	Contractor executes variations to lowest possible margin and never overorders materials and clients mostly do not pay reasonable variations in any event	Management	Client responsibility/communication/training
8	Through more efficient contractors	Contractors need to be more efficient and this will prevent variations	Management	Client responsibility/communication/training
9	This is facilitated by establishing good, interactive, regular communication with the client where programme and staging are discussed	Positive interactive relationships needed with client to focus on planning and scheduling	Management	Communications

(continued)

Table 7.32 (continued)

No.	Respondent comments to Question 32, Part 3 waste minimization	Coding	Theme	Sub-theme
10	Regular meeting with sub-contractors. Good central supervision	Regular sub-contractor meetings with central supervision running through project	Management	Communications/client responsibility
12	Work with client to minimize design changes during construction	Collaborative approach to minimize design changes	Design	Communications/FEED
13	Not as often as they should, however, under some contracts, say "cost plus" or alliances, there is often an incentive to do so	Appropriate contract models encourage variation minimization	Procurement	Communications/client responsibility
14	Yes, it is directly in regard to a module, process or work front. So, it is relevant to the individuals completing the tasks at management and work face	Informed management and workers need to interface with client on changes directly	Management/ workers	Communications
15	Depends very much on client/contractor relationships	Dependent on relationships between contractor and client	Management	Communications/training/client responsibility
16	It is not in contractor's interests to mitigate variations for commercial reasons	Contractors want to maximize variation costs	Management	Communications/client responsibility
17	Orders on need basis, minimizing volume of "consumables"	Order consumables on needs basis	Procurement	Quality process shortfall

By P. Rundle, February 2018

Table 7.33 Themes for waste minimization Part 3, Question 32

Theme No.	Theme description	Number of times this theme was recorded against respondent comment
1	Management	13
2	Procurement	6
3	Design	1
4	Workers	1

By P. Rundle, February 2018

Table 7.34 Sub-themes for waste minimization Part 3, Question 32

Sub-theme No.	Sub-theme description	Number of times this sub-theme was recorded against respondent comment	Remarks
1	Communications	15	
2	Client responsibility	7	One comment relates the importance of form of contract which is ultimately the client's choice
3	Training	6	
4	FEED	1	
5	Quality process shortfall	1	

By P. Rundle, February 2018

respondents believe "communication" between the contracted parties (15 respondent comments) was the key success factor in reducing site waste as a result of a client-initiated variation. The next two significant sub-themes were "client responsibility" for seven respondent replies and then "training" (six comments).

Fifteen respondents, by the very nature of their comments being from the contractor/builder respondent cadre, have made the telling point that effective communications between the client, the contractor/builder and other interested parties (i.e. design consultants, sub-contractors and vendors) are also critical in reducing the impact of variation scope changes or removing those changes all together, by inference, mitigating site waste.

The client has a duty of care to minimize this major potential source of waste by embarking on a strategy that ensures that scope changes will be kept to an absolute minimum (Larson and Gray 2011). Two key success factors in this strategy would be adequate pre-planning, including the development of a detailed scope of work and a valid user requirements specification, along with a set of contract documents that precisely set out the change order process and fairly apportion risk between the client and the contractor, along with the designer and the vendor (BHP Billiton 2007).

Table 7.35 Question 33, Part 3 waste minimization thematic analysis

No.	Respondent comments to Question 33, Part 3 waste minimization	Coding	Theme	Sub-theme
1	Is signed onto at the start of engineering	Waste minimization plan (WMP) adopted commencement design	Design	FEED
2	Generally, under our ISO 14001 EMS, these opportunities would be examined	WMP per code examined	Design	FEED/client responsibility
3	Yes—construction methodologies and planning are for time and cost reduction maximization leading to waste reduction	Construction methods based on cost reduction will reduce waste	Management	Client responsibility/training
4	Standard part of project plan prior to work commencement	WMP in project plan	Design	FEED
5	This is often addressed in our corporate management plans for sustainability, not sure if it is pushed down to the contractors	Client corporate sustainability plan contains WMP	Management	Client responsibility/communications
6	We minimize client waste whilst preparing estimate	Client waste reduced in estimates	Management	Communications
7	Working daily to minimize mistakes and forward thinking about each task	Contractor works daily to minimize error and plan ahead	Workers	Client responsibility/training
8	Commercial terms such as payment on surveyed installed quantities can help in this area, otherwise in a reimbursable contract only the client would be looking post hardening in the case of concrete, too late	Form of contract can influence site waste	Procurement	Client responsibility

(continued)

Table 7.35 (continued)

No.	Respondent comments to Question 33, Part 3 waste minimization	Coding	Theme	Sub-theme
9	Contractor is obliged to submit project management plan which includes environmental management plan	WMP is found in the environmental plan	Design	FEED
10	Typically, client driven as defined in the contract	WMP driven by client via contract	Management	Client responsibility, communication
12	Again, this is standard practice throughout the detail design and project scheduling activities	WMP is standard in design	Design	FEED
13	For each project, waste management plan is prepared to address relevant elements of the building and methodology of construction	Waste management plan is provided each contract	Design	FEED
14	Some do depends on project. Government agencies moving this way	Some projects, usually government have WMP	Management	Client responsibilit
15	Australian Capital Territory requires a project-specific waste management plan as part of project start-up	Australian Capital Territory government projects require waste management plan at mobilization	Design	FEED
16	A requirement in most infrastructure projects	WMP required most infrastructure projects	Design	FEED
17	Recycling targets included in project contract, and waste and recycling management plan in place including strategies for waste minimization	WMP included in waste recycling plan	Design	FEED
18	Categorization of potential waste and planning for segregation on site and responsible disposal/recycling	Projects have categorization potential waste for disposal/recycling purposes	Recycling	Recycling resour

By P. Rundle February 2018

Table 7.36 Themes for waste minimization Part 3, Question 33

Theme No.	Theme description	Number of times this theme was recorded against respondent comment
1	Design	9
2	Management	5
3	Procurement	1
4	Recycling	1
5	Workers	1

By P. Rundle, February 2018

7.3.1.7 Codes/Themes/Sub-themes for Respondent Comments for "Yes" Answer to Question 33, Part 3 Waste Minimization

Question 33—Interpretation of Key Findings

Question 33 asked respondents: *"does the contractor/builder and/or client have a mandatory waste minimization plan developed as part of the project execution plan?"*, to which the 48 respondents answered (Table 7.35):

  I. "No" for 33.33 per cent;
 II. "Don't know" for 29.17 per cent; and
III. "Yes" for 37.50 per cent.

Only 37.5 per cent of the 48 respondents to this question answered in the affirmative that their respective engineering construction companies had a site waste mitigation strategy in place. This was the equal lowest "yes" response; therefore, only slightly over a third of these respondent, companies have a waste reduction plan for their projects.

Refer to Table 7.36, which provides themes for this question developed after a thematic analysis and unequivocally highlights the respondents' opinion that "design" is the leading driver of a company site waste reduction plan with nine respondent comments, followed by five respondents claiming that "management" is a key driver, followed by one "procurement" comment. A review of Table 7.37, summarizing sub-themes for this question, indicates that the largest factor (nine comments) for this question on developing a site waste reduction plan, in the respondents' opinion, is the "FEED" sub-theme, followed by the "client responsibility" sub-theme with seven respondent replies.

Plainly, every construction project requires a basic waste minimization plan, as a precursor to a reduction process, and this requirement must become a mandatory project procedure. The only means to truly achieve this goal is to have this criterion become a client responsibility through the contract document as a condition of tender acceptance.

Table 7.37   Sub-themes for waste minimization Part 3, Question 33

Sub-theme No.	Sub-theme description	Number of times this sub-theme was recorded against respondent comment	Remarks
1	rccb	11	
2	Client responsibility	7	One comment makes the point of the importance of form of contract, which is ultimately a client responsibility
3	Communications	2	
4	Training	2	
5	Recycling resources	1	

By P. Rundle, February 2018

7.3.1.8   Codes/Themes/Sub-themes for Respondent Comments for "Yes" Answer to Question 34, Part 3 Waste Minimization

Question 34—Interpretation of Key Findings

Question 34 asked respondents if they could: "*provide other examples of situations where methods, processes or ideas were implemented on your construction site projects that minimized waste*", to which the 48 respondents answered:

  I. "No" for 16.67 per cent;
 II. "Don't know" for 33.33 per cent; and
III. "Yes" for 50.00 per cent.

Half of the 48 respondents of this research survey went to the effort of commenting on other possible means of reducing site construction waste on Australian projects, over and above those waste minimization strategies outlined in Questions 26–33, inclusive of this Part 3 survey questionnaire. Respondent comments were all constructive, but their remarks indicate that the abridged survey questionnaire pro forma, adapted from Yates (2013), provided a comprehensive suite of site waste minimization strategies. Table 7.38 exemplifies this conclusion, as the greatest number of respondent remarks was the "recycling" theme with ten comments, followed by "management" with nine respondent remarks, "procurement" with three comments and the "design" theme with three respondent comments (Table 7.39).

Table 7.40 supports this dearth of new site waste minimization strategies by noting that the "recycling resources" sub-theme, with seven respondent comments, was considered a key factor in "other" possible waste minimization strategies. Providing extra resources, more facilities and potential benefactors for recycled materials, is a fair strategy in considering a relentless drive to reduce construction waste.

Table 7.38 To Question 34, Part 3 waste minimization

No.	Respondent comments to Question 34, Part 3 waste minimization	Coding	Theme	Sub-theme
1	Establish community recycling groups to identify what materials can be salvaged and re-used beneficial to community	Use community as sources of beneficial waste re-use	Recycling	Recycling resources
2	Re-use of prohibited waste products such as contaminated soil or contaminated refractory bricks within the project boundaries, suitably contained via environmental controls	Maximize use of prohibited waste within site	Recycling	Innovative technical solutions
3	Water re-use	Recycle water	Recycling	Innovative technical solutions
4	Contractors scope to purchase/fabricate low-complexity/"small" items (e.g. piping and accessories <6")	Procure material to aid prefabrication	Procurement	Integrated supply chain
5	Large railed collapsible shutters/formworks that were re-usable for 1.5 km of lined tunnel	Purpose built repetitive formwork	Management	Innovative formwork solutions
6	Estimation of materials required, metal scrap bins provided, etc.	Tight procurement ordering/provide scrap bins	Procurement/recycling	Quality process shortfall/recycling resources
7	Redesigning to use structures that would otherwise be demolished	Use smart design to adapt existing structure	Design	FEED
8	Strong recycling strategies for waste materials	Develop detailed material recycling strategy	Recycling	Recycling resources
9	Tonkolli first-stage A$130 million project on green fields Africa. Documentation was solid	Develop strong contract documents	Procurement	Client responsibility

(continued)

Table 7.38 (continued)

No.	Respondent comments to Question 34, Part 3 waste minimization	Coding	Theme	Sub-theme
10	I have seen some waste separation on domestic residential sites and commercial residential projects, especially plaster board separation	Develop waste separation site operations	Recycling	Recycling resources
12	Develop waste management plan with recycling part	Proper waste and recycling plan	Management	Client responsibility
13	Known as professional project management	Rely on skilled project professionals	Management	Training
14	Once the contracts are let, nobody cares as long as there is a fully diverse waste receptacles for each form of waste on site. After that, everyone's obligations are generally dumped in the same common landfill	Prime concern is recycling waste bins and landfill access	Recycling	Recycling resources/client responsibility
15	Use of quarry as spoil dump or sediment disposal area	Use quarry as spoil dump site	Management	Innovative technical solutions
16	Sustainability focused contractual requirements including ISCA ratings and SDG minimum scores are often effective if the requirements are well written	Adopt prudent contract and implement through project starting with base design	Design/procurement	FEED/client responsibility
17	Minimizing waste is a constant on our projects. We do not always do it well, and we do not have systems/processes in place to monitor our waste minimization performance	Waste minimization is constant, but processes needed	Management	Client responsibility/training

(continued)

Table 7.38 (continued)

No.	Respondent comments to Question 34, Part 3 waste minimization	Coding	Theme	Sub-theme
18	Built our own concrete batching plant to manage issues with concrete for cast piles and bridge decks—was largely about waste	Optimize material control by running facility which minimizes waste issues	Management	Innovative technical solutions
19	Proactive policy that site waste is not buried on site and is re-used where possible	Proactive waste policy	Management	Training/recycling resources
20	Unpacking deliveries off-site and then delivered periodically to construction	Develop material packaging strategies to reduce site waste	Procurement	Innovative packaging solutions
21	A site manager must ensure that their foremen are on top of ordering concrete during concrete pours and volumes are correct as this is major waste product (excess concrete) which is expensive and difficult to cart off a constrained high-rise footprint	High-rise concrete ordering by fore-person is critical	Management	Quality process shortfall/training
22	Seeking resource recovery exemptions to maximize amount of material that can be re-used	Seek resource recovery exemptions	Recycling	Recycling resources
23	Paperless, engineering techniques	Adopt paperless office processes	Management	Innovative project management ICT support

By P. Rundle, February 2018

Table 7.39 Themes for waste minimization Part 3, Question 34

Theme No.	Theme description	Number of times this theme was recorded against respondent comment
1	Recycling	10
2	Management	9
3	Procurement	3
4	Design	2

By P. Rundle February 2018

Table 7.40 Sub-themes for waste minimization Part 3, Question 34

Sub-theme No.	Sub-theme description	Number of times this sub-theme was recorded against respondent comment	Remarks
1	Recycling resources	7	
2	Client responsibility	6	
3	Innovative technical solutions	4	
4	Training	4	
5	Innovative formwork solutions	2	
6	Integrated supply chain	2	
7	FEED	2	
8	Quality process shortfall	1	
9	Innovative project management ICT support	1	
10	Innovative packaging solutions	1	Respondent comments will be included in recommendations

By P. Rundle, February 2018

7.3.2 Summary of Results for Thematic Analysis of Front-End Site Waste Minimization Strategies

7.3.2.1 General Summary Thematic Analysis of Front-End Site Waste Minimization Strategies

A running summary of thematic analysis results for the front-end site waste minimization findings can be found at the conclusion of each of the sub-sections for Question 26 to Question 34, inclusive, in Sect. 7.3.1, "Thematic Analysis Findings

Table 7.41 Summary of Part 3, survey question content for front-end site waste minimization

Question No.	Part 3 survey questionnaire on site waste minimization strategies—Questions 26–32, inclusive
Question 26	Is your firm using techniques that improve resource efficiency, equipment efficiency and material resource efficiency and allow for training of manual labour?
Question 27	Are innovative designs, construction components, or construction processes, being integrated into your projects to reduced site generated waste?
Question 28	Do you adopt a structured approach both for engineering design and in the determination of construction methodologies that involve waste minimization strategies?
Question 29	Do you address waste generation reduction during project pre-planning to utilize designs that minimize waste using any of the following techniques: precast; prefabrication; pre-assembly or modularization?
Question 30	As builders, contractors and engineering consultants, do you ensure that a minimum amount of permanent and temporary materials are expended in the effective provision of client conforming construction/building products?
Question 31	Regarding the use of temporary construction materials, do you consider waste minimization processes? For example, for concrete construction, do designers specify concrete elements of similar dimensions, where practical; are steel shutters used on repetitive formwork; is formwork adequately treated and robustly fabricated, to allow re-use and are orders "just in time", to reduce material losses on site?
Question 32	Do the contractors/builders, consultants and vendors constructively work with the clients to minimize change orders that make pre-ordered products, redundant and suitable only for waste?
Question 33	Does the contractor/builder and/or client have a mandatory waste minimization plan developed as part of the project execution plan?
Question 34	Provide other examples of situations where methods, processes or ideas were implemented on your construction site projects that minimized waste.

By P. Rundle, February 2018

Comprising Coding/Themes/Sub-Themes for Part 3 Front-End Waste Minimization Strategies".

Note that Table 7.41 is provided as a ready reference for the reader to peruse in conjunction with the theme and sub-theme tables in this summary. Several of the questions are quite specific in their nature, such as Question 31, which highlights formwork and falsework, and Question 32, which discusses change order considerations.

Accordingly, the author has used due prudence when ranking themes and sub-themes in order of perceived importance, as per respondent comments. However, it is abundantly obvious that the three themes and four sub-themes that comprise the dominant factors in the thematic analysis are clear, via respondent comments, across most of the nine site waste minimization strategy survey questions.

Table 7.42 Summary of thematic analysis of themes for front-end waste minimization strategies

Themes	Number of respondent comments on a theme, as per question									Total
	Q26	Q27	Q28	Q29	Q30	Q31	Q32	Q33	Q34	
Management	10	15	5	17	8	10	13	5	9	92
Procurement	4	4	1	1	2	3	2	1	2	19
Design	3	6	14	4	4	6	1	9	2	49
Workers	1	–	–	–	–	2	1	1	–	5
Recycling	1	1	–	1	1	2	–	1	10	17

By P. Rundle, February 2018

Table 7.43 "Short-listed" critical themes for front-end waste minimization strategies

"Short-listed" critical themes for front-end waste minimization strategies	Total # of respondent comments, as per theme
Management	92
Design	49
Procurement	19

By P. Rundle, February 2018

7.3.2.2 Summary of Results for Themes Analysis of Front-End Site Waste Minimization Strategies

Table 7.42 provides a summary of the number of respondent comments for each theme for each of the nine questions, Questions 26–34, inclusive, on the research survey related to front-end site waste minimization strategies.

The following theme descriptions of all the considered themes, as per Table 7.43, comprehensively detail exactly how the critical themes "short-list" were developed for these themes. Theme descriptions used in this analysis are noted below.

"Management" Theme

The "management" theme refers to the synthesis of respondent data using thematic analysis to determine codes and themes for Part 3, Questions 26–34, inclusive, in this chapter, "Findings", covering front-end site waste minimization strategies. The "management" theme is the major factor impacting site waste reduction at the front-end of the physical site waste cycle. The "management" theme was developed from coding of participant remarks and factors identified, including such issues as a need for good planning; the provision for the project of the latest BIM engineering and paperless document control ICT systems; a management structure that has the foresight to plan projects from inception, using documents that encourage site waste minimization from the initial design phase; ensuring that the client, contractor and designer all work to a common waste reduction management plan; considering incentives for site waste reduction and profit sharing of these value-adding initiatives; and involving contractors at an early stage of design. In summary, the survey respondent sample pool, which comprises members of the engineering construction management profession in various fields including engineering, contractor, field supervision, client, technical and project management fields, has overwhelmingly nominated the "management" cadre as the most important theme related to site waste minimization, across Questions 26–34, inclusive.

"Design" Theme

The "design" theme was the next most important theme in creating sources of site waste. The predominant code for this theme was a requirement to undertake adequate preliminary design and project development as early as possible. Other coding referred to a need to design, where appropriate, a system which allowed steel structures to be modularized off-site for the implementation of value engineering workshops and to ensure that the design specifications account for construction waste mitigation.

"Procurement" Theme

The third highest theme was determined as "procurement", impacting on front-end site waste mitigation strategies. Codes variously identified procurement errors in areas such as poor material control (including overordering, incorrect material take-offs and other ordering errors), poor logistics (including improper material handling and poor storage facilities) and improper handling of materials at the work face.

However, the predominant re-occurring code in around half of the "procurement" themes synthesized from respondent comments on site waste minimization was the requirement for an integrated materials supply chain, as a process to drive project materials flow as an important potential waste reduction control tool.

"Workers" Theme

There were only five comments regarding the "workers" theme in all respondent comments on Questions 26–34, inclusive, for site waste minimization strategies, compared to 92 comments for the "management" theme. As the number of respondent remarks was only five for this theme, it was not "short-listed" for further analysis.

"Recycling" Theme

There were 17 respondent comments on "recycling", ten of which were captured in Question 34, which asked for other possible front-end waste minimization strategies, not noted in Questions 26–33, inclusive, of this research survey questionnaire. Several proactive comments on the sources of site waste contained remarks on how to make recycling more efficient by the provision of adequate personnel and facilities resources up front for recycling, and these remarks are commented on elsewhere. However, as this downstream recycling/material re-use set of options was specifically excluded from this study and noted as such in the survey questionnaire, the "recycling" theme was not "short-listed" for further review.

7.3.2.3 Summary of Results for Sub-themes Analysis of Front-End Site Waste Minimization Strategies

Table 7.44 provides a summary of the number of respondent comments for each sub-theme for each of the nine questions in this research survey related to front-end site waste minimization strategies.

Upon mature reflection of all of the available sub-theme data summarized in Table 7.44, the author adopted extensive sub-themes, due to the varied nature of the nine questions surrounding front-end site waste minimization and more particularly the broad spectrum of comments received from respondents. Accordingly, it has been deemed practical to consolidate the four sub-themes that refer to innovation, namely innovative project management ICT support; innovative technical solutions; innovative formwork solutions; and innovative packaging solutions into a single "innovation" sub-theme.

This consolidated sub-theme appears in Table 7.45, which summarizes the "short-list" of critical sub-themes. This is denoted as the "innovation sub-theme for the following: project management ICT support; technical solutions; formwork solutions; and packaging solutions".

All of the sub-themes prepared in this analysis and summarized in Table 7.44 are described below. These following sub-theme descriptions fully detail exactly how Table 7.45 was developed for these themes. Theme descriptions used in this analysis are noted below.

"FEED" Sub-Theme

"Front-end engineering design" (FEED), as pointed out by respondent survey comments, is the most critical sub-theme. "FEED" is a specific design process that addresses the criteria under which a design must be executed from engineering inception, with a view to minimizing site waste, which would be a mandatory contract requirement (Yates 2013). Using this approach, the client can maximize the opportunities to reduce site waste by ensuring the prime construction, supply, vendor contracts and consultant design contracts guarantee that a "universal waste management plan" is included in all such project execution contracts (Udawatta et al. 2015). Accordingly, "FEED" is taken to mean all early project management, design, specification and contract document preparation in this context.

"Innovation" Sub-theme for the Following: Project Management ICT Support; Technical Solutions; Formwork Solutions; and Packaging Solutions

A bracket of four innovation sub-themes has been consolidated into one item, "innovation", as the next most critical sub-theme after "FEED". Myers and Marquis (1969) suggest that "innovation" is the initial commercial transaction of a new technological process. Slaughter (1998) classified construction innovation typology as incremental, radical, architectural, modular and system. Murphy et al. (2008, p 102) states that "Slaughter's (1998) examples of these innovations were drawn primarily from civil engineering projects. The innovations contributed to the buildability and functionality of a building, e.g. [sic] on-site construction processes". Blayse and Manley (2004) note that the Slaughter model (1998) definition on construction innovation

Table 7.44 Summary of thematic analysis of sub-themes for front-end waste minimization strategies

Sub-theme	Number of respondent comments on a sub-theme, as per question									Total
	Q26	Q27	Q28	Q29	Q30	Q31	Q32	Q33	Q34	
Innovative project management ICT support	4	–	–	–	–	3	–	–	1	8
FEED	3	3	13	4	4	6	1	9	2	45
Integrated supply chain	3	–	2	–	–	2	–	–	2	9
Work culture	3	–	–	–	–	–	–	–	–	3
Client responsibility	2	2	2	1	3	3	7	7	6	33
Quality process shortfall	1	–	–	–	–	–	1	–	1	3
Maximize off-site construction fabrication and pre-assembly	–	6	1	17	1	2	–	–	–	27
Innovative technical solutions	–	5	2	–	–	–	–	–	4	11
Innovative formwork solutions	–	1	–	1	–	8	–	–	2	12
Innovative packaging solutions	–	2	–	1	–	–	–	–	–	3
Contractor presumption of who pays	–	–	–	–	5	–	–	–	–	5
Recycling resources	–	–	–	–	1	2	–	2	7	12
Communications	–	–	–	–	–	–	15	2	–	17
Training	3	–	–	–	1	2	6	2	4	18

By P. Rundle, March 2018

Table 7.45 "Short-listed" critical sub-themes for front-end waste minimization strategies

"Short-listed" critical sub-themes for front-end waste minimization strategies	Total # of respondent comments, as per sub-theme
FEED	45
Innovation for following: project management ICT support; technical solutions; formwork solutions; and packaging solutions	34
Client responsibility	33
Maximize off-site construction fabrication and pre-assembly	27

By P. Rundle, March 2018

is the generally accepted description. Slaughter (1993, p. 83) defines innovation as follows:

> innovation is the actual use of a nontrivial change and improvement in a process, product, or system that is novel to the institution developing the change. Innovation in the construction industry can take many forms.

There has been a real drive by several of the survey respondents in each of their particular sectors (planning and scheduling; cost estimating; field project management; and engineering design), to stay abreast of the latest available information communications technology (ICT) as tools for each of their respective fields to assist in project management activities effectively and efficiently and as a flow on benefit assist with front-end waste reduction.

Similarly, respondent comments to Part 3 of this survey questionnaire, on front-end site waste minimization, indicate the need to develop unique innovative solutions to waste reduction (whatever that innovation may be—in the design area, particularly). There is a general sense, in reading the Question 26 to Question 34, inclusive, respondent comments in Part 3 survey on site waste minimization that there is a decided emphasis on a willingness by contractors/builders to keep abreast of concrete formwork/falsework innovation. Question 31 received the majority of these responses on formwork/falsework, with these temporary work elements used as an example in framing the survey question.

There were several comments on packaging and pallet site waste issues, and the need for innovation when working with suppliers to develop better solutions and these comments shall be included in Sect. 8.7, "Packaging Material Site Waste Survey Discussion".

"Client Responsibility" Sub-theme

The "client responsibility" sub-theme was the third most critical sub-theme, as per Table 7.45. Refer to Sect. 7.2.4.4, "Sub-themes Developed, as per Question 25 Respondent Comments on Sources of Site Waste", which discusses the factors that are considered by the "client responsibility" sub-theme.

"Maximize Off-site Construction Fabrication and Pre-assembly" Sub-theme
This is the fourth and final critical sub-theme, and it relates to fabricating construction elements, such as pipework and structures, off-site. It also includes pre-assembly of permanent material elements, such as process plant modules and electrical sub-stations. Work is executed in a controlled environment, which greatly increases the opportunity to minimize waste (Gibb 1999). However, there can be extra-over material utilization in process module pre-assembly to allow for transportation and materials handling (Gibb 1999). Notwithstanding this, Jaillon et al. (2009) convincingly argues that pre-assembly processes can affect a savings in material waste of approximately 52 per cent. There is a strong thread in respondent comments supporting the use of precast, for example, on high-rise commercial buildings and pre-assembly, modularization and prefabrication on heavy construction projects. However, there are no noted comments on residential work.

The author posits, anecdotally yet from direct experience, that opportunities for the use of precast concrete versus in situ concrete construction; modularization of process elements versus stick build and fabricating pipework spools; and off-site versus in situ, are all important construction methodology decisions that must be made at the very early stage of a project in close consultation with the design consultants. These outcomes will be determined by many factors, such as the remoteness of the environment; site crane access and crane size availability; the quality of site and off-site labour; the industrial relations impacting any advantages in off-site fabrication; transportation access; and, for example, statutory approval implications for overseas prefabrication. There are many good examples of extensive use of prefabricated structures in remote environments, particularly in Australia, such as the prefabricated airport terminal and office complex at the Santos Gas facility in Ballera, Queensland, and the prefabricated tract housing for BHP Billiton Mitsubishi in Moranbah, Queensland.

Chandler (2016) suggests that as little as five per cent of building work in Australia uses elements of prefabrication. This is an area that needs further investigation to facilitate a reduction in site waste. There are many elements in commercial building and, particularly, residential building, such as wall frame and roof truss prefabrication, that reduce costs and minimize material waste (Doab 2018). Further research and marketing of these developments need to be undertaken on these prefabrication costs.

"Integrated Supply Chain" Sub-theme
Since the mid-1990s, the international construction industry has been reviewing the benefits of moving towards an integrated supply chain management process, with varying success, as a means of replicating this effective and efficient process adopted in other industries, such as production engineering (Khalfan et al. 2004). However, Green and May (2003) suggest that the unique nature of the construction industry, and the non-repetitive nature of many of the work activities, makes a production-type process untenable. In response to the different types of work presented in the construction business, abridged models have been generated that emphasize the integration between all suppliers and a requirement to work together collaboratively and reduce

conflict potential (Khalfan et al. 2004). Notwithstanding the arguments for/against this approach, the main driver has been to develop a process that links suppliers into upstream and downstream chains, with the aim of producing a quality optimized product by integrating supplier processes and activities (Khalfan et al. 2004). This is an important sub-theme and will be remarked upon in Chap. 8, "Discussion", because nine of the 19 procurement theme remarks concerned an integrated chain sub-theme. Notwithstanding this, it is not considered a critical sub-theme.

"Training" Sub-theme

Training to effect cultural change in the Australian engineering construction industry for manual workers and supervisory, technical and management staff is critical to developing a mentality of site waste reduction (Lewin 1947). This is not considered a critical sub-theme, as per the comments and the number of participant responses. However, to successfully undertake the difficult management task of bringing about an industry-wide change in cultural practices of around one million Australian engineering construction employees to focus on front-end site waste reduction, continual training will be required to inculcate and prevent reversion from these new desired behaviours (Thompson et al. 2014).

"Contractor Presumption of Who Pays" Sub-theme

The author has read a certain hubris into the comments received from several contractor/builder respondents replying to Question 30. The author, after due self-reflection, asks whether these responses may indicate that some parties believe that the client ultimately pays for wasted materials as part of the contract sum. This is not considered a critical sub-theme, as per the respondent comments and the number of participant responses.

"Communications" Sub-theme

The "communications" sub-theme was discussed mainly in the Question 32 comments from respondents, where it dominated with 15 participant remarks. This question surrounded the constructive development of change orders and working together with the client to reduce site waste, because of scope changes potentially making materials redundant. Jaffar et al. (2011) highlights the need to maintain open lines of communication between the contractor/builder, designer and client when handling variation changes to project scope and of having set conflict resolution processes in place to negotiate outcomes constructively and equitably. This sub-theme is prescient on several counts. Firstly, client-initiated design changes have been highlighted, both in this research and by triangulation with other comparative past academic studies (Faniran and Caban 1998; Nagapan et al. 2012; Yates 2013), as a leading source of construction site waste. Secondly, *prima facie*, there appears to be a *laissez-faire* attitude in some contractor/builder comments that whatever the material costs are, they will simply be passed on to the client. Finally, the site waste minimization component of this study has clearly shown that there is a trichotomy of differing and often conflicting interests, between the client, designer and builder/contractor surrounding change orders, due to reasons of commercial profit. "Communication" is always an important factor in all forms of project management, but is not considered a critical sub-theme, as per the respondent comments and the number of participant responses. However, it will be remarked upon in Chap. 8, "Discussion".

"Quality Process Shortfall" Sub-theme

The reader's attention is drawn to Table 7.9 of Sect. 7.2.4.4, "Sub-themes Developed, as per Question 25 Respondent Comments on Sources of Site Waste", which shows that the "quality process shortfall" theme is the most critical sub-theme for potential sources of construction waste. However, Table 7.44, "Summary of Thematic Analysis of Sub-themes for Front-End Waste Minimization Strategies", transparently indicates that, with only three sub-themes coded to respondent comments, it is not a critical sub-theme for front-end site waste minimization strategies.

The Faniran and Caban (1998) site waste identification model was very detailed in its list of possible sources of waste. The author viewed this as a "tactical" document that could be used in the field as a checklist on a site waste process audit, for example. However, the adapted Yates (2013) site waste strategies model represented broad strategies. Hence, Sect. 7.2, "Major Sources of Site Waste Findings", can be viewed as a process driven view, whilst the Sect. 7.3, "Font-End Site Waste Minimization Strategy Findings", is simply defined as a high-level plan for reaching specific business objectives (Thompson et al. 2014). Therefore, the importance of any quality control process "shortfalls" for Question 25 was evaluated by thematic analysis for sources of waste, compared to the nine-question investigation of site waste minimization strategies by thematic analysis (Questions 26–34, inclusive).

The "quality process shortfall" sub-theme encompasses project management processes that have not been followed and/or project management work activities that require a process to be prepared and implemented. This sub-theme has been prepared using the rationale that if a quality assurance audit was undertaken, then a shortfall (i.e. non-compliance) would uncover the waste source causation noted by the respondent in their remarks on the sources of site waste. For example, defective material quality, poor material handling, incorrect material storage, overordering, incorrect ordering, poor-quality material and sub-standard workmanship are all quality assurance shortfalls (i.e. process non-compliances) in the project management system. Sub-standard supervisory staff and poor project planning are also considered quality non-compliances.

This theme has not been considered a critical sub-theme, as per the above comments and the number of participant responses for this Sect. 7.3, "Front-End Site Waste Minimization Strategy Findings".

"Work Culture" Sub-theme

"Work culture" received several comments in the sources of waste thematic analysis contained in Question 25, earlier in this Sect. 7.3.2.3. However, it has not been considered a critical sub-theme, as per the comments and the number of participant responses for this waste minimization strategies thematic analysis sub-section.

Notwithstanding this, what will be discussed further is the very successful change in the engineering construction and broader community work culture via the acceptance of recycling practices that have been introduced, reinforced and maintained through a tripartite approach of "a winning of community hearts and minds" environmental campaign, government regulation and enforcement by punitive legislation, to make recycling of site waste products the standard practice internationally, particularly in the developed world economies (Chandler 2016; Hickey 2015; Ritchie 2017).

Recycling Resources Sub-theme

There were several respondent comments that alluded to recycling efforts being hampered at the front-end of the construction waste cycle by (i) a lack of experienced personnel to grade construction waste products for potential recycling re-use; (ii) inadequate number of recycling bins strategically placed around the site; (iii) insufficient personnel to recycle into bins on site; and (iv) an insufficient laydown area to be allocated for recycling. However, these constructive comments were mirrored in the earlier thematic analysis, in this Sect. 7.3, addressing Question 25 on "sources of site waste". The majority of comments on recycling regarding front-end waste minimization strategies covered in Question 34 of this research survey were off-topic and had insufficient numbers of respondent comments across the survey to warrant being nominated as a critical sub-theme.

7.3.3   Summary of Findings for Analysis of Respondent Responses on Front-End Waste Minimization Strategies, as Per Yates (2013)

The Yates (2013) paper (that Part 3 survey questionnaire on front-end waste minimization strategies for this research used in adapted form) did not use thematic analysis to synthesize the data results. Each of the strategies was evaluated and commented on with respect to respondent comments. Accordingly, this Sect. 7.3 has followed this approach and has undertaken the detailed thematic analysis, as summarized in Sect. 3.3.2, "List Peer Review Panel".

7.3.3.1   Summary of Findings for Analysis of Respondent Responses on Front-End Waste Minimization Strategies

Question 26 of Part 3 of this research survey asked respondents to comment on how efficient their firms are with personnel, resources and equipment use. The "yes" response of only 41.7 per cent seemed very low to the author. Salient points are found in Table 7.46, "Discussion Respondent Answers—Part 3 Site Waste Minimization Strategies Question 26 to Question 34", in the right-hand side column denoted "key findings". It is worth reiterating the following major findings on front-end waste minimization strategies.

This review of Question 33 establishes that there appears to be a low number of respondents who are working in firms with a mandatory site waste reduction plan as part of their project execution plan. Question 34 asked respondents to comment on

Table 7.46 Discussion respondent answers—Part 3 site waste minimization strategies Question 26—Question 34

Q. No.	Question description	Respondent answered "no" %	Respondent answered "don't know" %	Respondent answered "yes" %	Key findings
26	Is your firm using techniques that improve resources efficiency, equipment efficiency and material resource efficiency and allow for training of manual labour?	39.58	18.75	41.67	This is quite a low score of 41.7 per cent respondent affirmative replies, considering benefits that successful strategy outcome can provide
27	Are innovative designs, construction components or construction processes, being integrated into your projects to reduce site generated waste?	31.25	20.83	47.92	Only 37.5 per cent of respondents stated that their company had a mandatory waste minimization plan, but nearly 48 per cent answered "yes" for this question, which indicates that the experience of engineering manager and project manager in planning the work is critical to waste reduction considerations—rather than mandatory waste plan

(continued)

Table 7.46 (continued)

Q. No.	Question description	Respondent answered "no" %	Respondent answered "don't know" %	Respondent answered "yes" %	Key findings
28	Do you adopt a structured approach both for engineering design and in determination of construction methodologies that involve waste minimization strategies?	29.17	20.83	50.00	Only 37.5 per cent of respondents stated that their company had a mandatory waste minimization plan, but nearly 50 per cent answered "yes" for this question, which indicate that the experience of engineering manager and project manager in planning the work is critical to waste reduction considerations—rather than mandatory waste plan
29	Do you address waste generation reduction during project pre-planning to utilize designs that minimize waste using any of the following techniques: precast; prefabrication; pre-assembly or modularization?	29.17	14.58	56.25	These construction practices noted reduce both labour and material costs working in controlled, off-site environments. Up early design is necessary to allow for such waste reduction construction methodologies. Also, important is early contractor involvement design and other practices such value engineering workshops at concept design stage

(continued)

Table 7.46 (continued)

Q. No.	Question description	Respondent answered "no" %	Respondent answered "don't know" %	Respondent answered "yes" %	Key findings
30	As builders, contractors and engineering consultants, do you ensure a minimum amount of permanent and temporary materials are expended in the effective provision of client conforming construction/building product	31.25	31.25	37.50	Because of the high unit cost of Australian labour, there is a greater focus on effective and efficient use of personnel and if that means using more material, contractor/builder will do so. Refer to Question 29 and Question 31 and note this high "yes" score indicates the need to drive down labour costs by working efficiently and effectively
31	Regarding use of temporary construction materials, do you consider waste minimization processes? As an example, for concrete construction, do designers specify concrete elements of similar dimensions where practical? Are steel shutters used on repetitive formwork; is formwork adequately treated and robustly fabricated, to allow re-use and are orders "just in time" to reduce material losses on site?	18.75	29.17	52.08	These construction practices noted reduce both labour and material costs working in controlled, usually, off-site environments

(continued)

Table 7.46 (continued)

Q. No.	Question description	Respondent answered "no" %	Respondent answered "don't know" %	Respondent answered "yes" %	Key findings
32	Does the contractor/builder, consultants and vendors, constructively work with the client to minimize change orders that make pre-ordered products, redundant and suitable only for waste?	35.42	22.92	41.67	The respondent comments in Part 3 questionnaire ⊂ Question 32, and other questions show there is a trichotomy of potentially conflicting interests between client, contractor/builder and consultant, surrounding profit making on client change orders
33	Does the contractor/builder and/or client have a mandatory waste minimization plan developed as part of the project execution plan?	33.33	29.17	37.50	Only 37.5 per cent of respondents are certain that their firm has a mandatory waste minimization plan, and 33.3 per cent of respondents are sure there is no plan, with the balance of respondents unsure

(continued)

Table 7.46 (continued)

Q. No.	Question description	Respondent answered "no" %	Respondent answered "don't know" %	Respondent answered "yes" %	Key findings
34.	Provide other examples of situations where methods, processes or ideas were implemented on your construction site projects that minimized waste	16.67	33.33	50.00	A review of respondent comments for this question indicates that there were a number of valuable remarks on site waste minimization, but these responses were mostly variations of the strategies listed in this table that appeared on the survey questionnaire. However, the most common comment pertained to "recycling", which the questionnaire asked respondents to preclude, as the research targeted front-end site waste reduction. Ten of the 24 respondent themes in Question 34 addressed recycling (see Table 7.39)
Total respondent answers		264.59	220.82	414.59	
Average for each respondent answer		29.40% = "no"	24.54% = "don't know"	46.07% = "yes"	

By P. Rundle, March 2018

other front-end strategies that could reduce site waste, over and above the strategies nominated in the research Part 3 questionnaire. The research survey specifically precluded any remarks on downstream strategies including "recycling", most of the responses mentioned recycling as an option. Ten of the 24 respondent themes in Question 34 addressed recycling (see Table 7.39). However, it was encouraging to see the considerable extent to which the engineering construction community has been effectively indoctrinated to consider recycling of waste as the ultimate solution for waste reduction, as can be seen by this summary. Notwithstanding this, a cultural change needs to be augmented to alert both academics and professionals to the fact that it is more beneficial to key stakeholders, the community at large and the environment to reduce, and preferably entirely remove, site waste potential, than to handle a downstream waste product via recycling/re-use. Since saturation has been reached, as per Question 34 results, no new front-end strategies were outlined via respondent comments.

The author has been best able to address respondent outcomes for Questions 26–34, inclusive, using the academic literature, including similar research that used comparable models (Faniran and Caban 1998; Nagapan et al. 2012; Yates 2013). Refer to the next Sect. 7.3.3.2, "Summary of Findings for Analysis of Respondent Responses on Front-End Waste Minimization Strategies Compared to the Academic Literature", for further analysis of research data, via triangulation.

7.3.3.2 Summary of Findings for Analysis of Respondent Responses on Front-End Waste Minimization Strategies Compared to the Academic Literature

The author was looking for a sense of what the average percentage of respondents answering in the affirmative "yes" may be viewed as a valid response for, as opposed to "no" or "don't know". Admittedly, the Yates (2013) survey questionnaire, used to identify site waste reduction strategies, was slightly adapted in the author's research to delete any "back-end recycling questions" and replace them with a question on "other possible strategies". However, the five core questions are the same. Interestingly, an analysis of both this textbook and the Yates (2013) data showed that both surveys attained a 46 per cent "yes" vote as an average. Though due prudence needs to be given to this comparison, it is nevertheless a useful evaluation. Table 7.47 compares this book's research findings with the Yates (2013) model for site waste minimization strategies.

Udawatta et al. (2015) remarked on the high number of contractor/builder participants who did not have a specific project waste plan in the landmark qualitative research undertaken by Faniran and Caban (1998) on sources of site waste on Australian projects (Udawatta et al. 2015). Their survey stated that 42.9 per cent of respondent companies did not have a site waste reduction policy, whilst 57.1 per cent of the respondents answered in the affirmative (Faniran and Caban 1998). In reference to more recent data from the USA, the Yates (2013) paper, as cited in

Table 7.47 Comparison of site waste minimization research and Yates (2013) findings

Question no.	Question description	This research respondent answers to Part C survey			Yates (2013) research data using same survey questions		
		"No" %	"Don't know" %	"Yes" %	Respondent answered "no" %	Respondent answered "don't know" %	Respondent answered "yes" %
26	Is your firm using techniques that improve resource efficiency, equipment efficiency and material resource efficiency and allow for training of manual labour?	39.58	18.75	41.67	26	29	56
27	Are innovative designs, construction components, or construction processes being integrated into your projects to reduce site generated waste?	31.25	20.83	47.92	19	42	39
28	Do you adopt a structured approach both for engineering design and in the determination of construction methodologies that involve waste minimization strategies?	29.17	20.83	50.00	23	19	58
33	Does the contractor/builder and/or client have a mandatory waste minimization plan developed as part of the project execution plan?	62.5 per cent did not know and did not have mandatory waste management plan		37.50	50 per cent did not know and did not have mandatory waste management plan		50

By P. Rundle, March 2018 (adapted from Yates 2013)

Table 7.47, states that 50 per cent of respondents in the qualitative survey noted that their company had a waste management plan.

Pursuant to the above, the 37.5 per cent of respondents in this research who answered in the affirmative to having a mandatory waste minimization plan appears very low. Considering several of the other questions pertaining to following a standardized approach to design/construction methodologies in endeavouring to reduce site waste, the data presented in this textbook indicates that it is in closer alignment with the Yates (2013) data synthesis of respondent comments on site waste minimization; Table 7.47 refers in detail. A review of Udawatta et al. (2015) may explain the low respondent answer in this research of only 37.5 per cent of participants answering "yes" to their firm having a mandatory waste minimization plan. Table 4 of the Udawatta et al. (2015) paper notes that their quantitative research showed that their Australian participants rated "good company policies on construction waste management" only tenth, whilst the incorporation of waste management procedures in early design was noted as the fourth most important waste reduction strategy. This could give the impression that there are waste reduction measures being adopted, but not necessarily formalized, into company standard operating procedures.

The results in Table 7.47 indicate that there is a possibility that there is more of an emphasis on innovation in the Australian construction industry as a way to reduce site waste than in the USA, as the Yates (2013) USA data points to. It is necessary to bear in mind that construction and demolition statistics indicated for both Australia and the USA are very similar—around 40 per cent of landfill in both countries comprises construction and demolition waste (Ritchie 2016).

Table 7.47 comparison between this research and the USA results from Yates (2013) indicates that there is more of an emphasis on personnel, equipment and resource efficiencies with USA contractors/commercial builders, in comparison to Australian counterparts. In fact, academic literature from Langston (2012) shows that USA projects are, on average, 37 per cent more efficiently constructed than similar Australian projects.

7.3.4 Key Points from the Author's Reflexivity Journal for Respondent Comments on Site Waste Minimization Strategies from Questions 26 to 34, Inclusive, for Thematic Analysis Process

Table 7.48 provides key factors from the author's reflexivity journal that progressively logged reflections on emerging patterns, themes and concepts during the data analysis for the nominated respondent comments, as per Questions 26–34, inclusive, of Part 3 of the survey questionnaire on waste minimization strategies. These reflections were progressively recorded from the coding stage, right through to the logging of any changes to themes/sub-themes during the review phase (Saldana 2009).

Table 7.48 Reflexivity journal extracts on Part 3 survey on site waste minimization

No.	Respondent opinions on possible waste minimization strategies	Key points from the author's reflexivity journal whilst developing codes/themes/sub-themes
Q26	Is your firm using techniques that improve resources efficiency, equipment efficiency, material resource efficiency and allow for the training of manual labour?	
Pt15[a]	We are designers and site supervisors and do not have a direct effect on construction materials procurement, delivery and utilization/construction, to manage or affect contractor waste	There seems to be a degree of intransigence with client, design consultant and contractor, about which entity takes responsibility (i.e. the risk) for site waste reduction. One needs to avoid "taking sides" or being dismissive when reviewing the area of risk apportionment. The author needs to be free of any preconceived bias he has on this matter
Q27	Are innovative designs, construction components, or construction processes, being integrated into your projects to reduced site generated waste?	
Pt2	Promoting single package bundling of multiple components rather than multiple, individual wrapped components. Delivery and interim storage in containers to provide protection until use—results in reduction of weather proof packaging	This is a positive example of a proactive respondent comment on what survey participants see as a significant source of waste. It is a lesson in focusing on a problem solution. The author needs to focus on the positives as well as the negatives of respondent comments
Pt5	Yes, but not directly to reduce waste but to provide an efficient design. It is assumed that the contractor will manage waste as this would reflect in construction cost and time reductions	There seems to be a silo raised by designer, here, only being responsible for an efficient design that would/could reduce waste, rather than a constructive early involvement approach, working with contractor. The author needs to be free of any preconceived bias he has on this matter
Pt20	As per above, innovative design solutions for us means "smarter and easier" construction in the field and this reduces waste through error in a tough and competitive commercial building sector in Melbourne	This is a "tight" comment from a professional high-rise project site manager recognizing importance of sharp design which mitigates building errors and hence reduces waste. This context needs to come through in the recommendations of textbook

(continued)

Table 7.48 (continued)

No.	Respondent opinions on possible waste minimization strategies	Key points from the author's reflexivity journal whilst developing codes/themes/sub-themes
Q28	Do you adopt a structured approach both for engineering design and in the determination of construction methodologies that limit waste minimization strategies?	
Pt9	For design and build contracts where there is economic incentive to design out waste or in response to sustainability rating schemes (green building/infrastructure)	This comment points to a need for cultural shift in the workplace; namely, all projects should design out waste not just C&D where contractor/builder would (probably) profit the most. The author needs to be free of any preconceived bias he has on this matter
Q29	Do you address waste generation reduction during project pre-planning to utilize designs that minimize waste using any of the following techniques: precast; prefabrication; pre-assembly or modularization?	
Pt17	Most projects push the responsibility of waste management to EPC or EPCM who in turn push it further down the commercial pecking order and finally to landfill	The author needs to reflect on whether he is showing "bias" towards risk aspects of the waste reduction problem, or these recurring comments from client, contractor and design consultant all appear to abrogate primary responsibility for waste management
Pt23	Assessing the design to suit logistics (roads transport/lifting capacities), implementation of off-site manufacturing, off-site partial pre-assembly	Prudent comment on practical design/logistics interface, as per large-scale prefabrication and impact on existing bridge loadings, for example
Q30	As builders, contractors and engineering consultants, do you ensure a minimum amount of permanent and temporary materials are expended in the effective provision of client conforming construction/building product?	
Pt7	If we do not do this, then we do not win work	*Prima facie*, this is a "comfortable" comment which may or may not be rooted in the ultimate assumption that the owner will pay for the total contract sum including any generous heuristic (or actual) cost allowances for waste. The author needs to be free of any preconceived bias he has on this matter
Pt16	Where there is opportunity through the contract and mutual benefit (e.g. green building)	A cultural change required, no doubt from the source of funding—the client, to ensure site waste reduction is always considered on projects. The author needs to be free of any preconceived bias he has on this matter. This author assertion will need to be supported by academic and/or professional literature

(continued)

Table 7.48 (continued)

No.	Respondent opinions on possible waste minimization strategies	Key points from the author's reflexivity journal whilst developing codes/themes/sub-themes
Pt17	We deliver to the client a healthy standard of workmanship and minimal wastage to minimize actual cost	It is certain that the author of these comments believes that this is occurring and no doubt, it is. However, only around half of the company employees polled believe waste minimization strategies are in place for each of their companies. More needs to be done to formulate firm policies on waste reduction
Q31	Regarding the use of temporary construction materials, do you consider waste minimization processes? For example, for concrete construction, do designers specify concrete elements of similar dimensions where practical; are steel shutters used on repetitive formwork; is formwork adequately treated and robustly fabricated, to allow re-use and are orders "just in time", to reduce material losses on site?	
Pt12	Standardization is virtually non-existent. Precast is almost an evil conjuring. How could the large engineering houses make profit from cookie cutter approach. The baulk at pre-assembly because it minimizes post-design-phase follow-up works	These forceful respondent comments lead the author to believe there is an element/degree of mistrust between the various sectors within the Australian engineering construction industry between principal, engineering consultant and contractor/builder regarding their individual intent beyond readily maximizing profit margin. The author wishes to work with this basic business desire to make the necessary cultural shift
Pt15	As a PMC, these are not directly controlled or specified in contracts. However, selected sub-contractors (e.g. concrete construction) do make use of re-usable formwork	In Question 29, Pt 17, above, the client was pushing responsibility onto their EPCM contractor. Now, the EPCM wants to hand this responsibility to their sub-contractors. The author needs to be free of any preconceived bias he has on this matter
Q32	Do the contractors/builders, consultants and vendors constructively work with the client to minimize change orders that make pre-ordered products, redundant and suitable only for waste?	
Pt5	"Managing the client" is still a challenge	There is a trichotomy of interests between owner, contractor and consultant regarding the variations. The author needs to be free of any preconceived bias he has on this matter about which party is in the right—it is a case by case issue and "ditto" for Pt 7, 8 and 16, below, per this question
Pt7	Any variation has to be carried out to lowest cost. The contractor does not overorder, as the client usually will not pay reasonable costs in every case	There is a trichotomy of interests between owner, contractor and consultant regarding the variations

(continued)

Table 7.48 (continued)

No.	Respondent opinions on possible waste minimization strategies	Key points from the author's reflexivity journal whilst developing codes/themes/sub-themes
Pt8	Through more film tin contractors	There is a trichotomy of potentially conflicting interests between owner, contractor and consultant regarding the variations
Pt16	It is not in contractor's interests to mitigate variations for commercial reasons	There is a trichotomy of potentially conflicting interests between owner, contractor and consultant regarding the variations
Q33	Does the contractor/builder and/or client have a mandatory waste minimization plan developed as part of the project execution plan?	
Pt5	This is often addressed in our corporate management plans for sustainability, not sure if it is pushed down to the contractor	This is a comment from a major capital works client which again indicates the need to address the issue of apportionment of responsibility for site waste reduction on projects. The author needs to be free of any preconceived bias he has on this matter
Pt6	We minimize client waste whilst preparing estimate	The author contacted an estimating firm about this comment, and the company asserted that estimators set sensible limits of waste that are transparent in their estimate for all major commodities, to keep waste volume manageable
Pt10	Typically, client driven as defined in the contract	Suggest that this is the appropriate response with the caveat that the client contract covering site waste minimization plan be a regulatory codified document. The author leans towards this approach after reviewing the results and his contention will need to be evaluated by further literature review
Pt15	Australian Capital Territory requires a project-specific waste management Plan as part of project start-up/	The approach described in this respondent comment is *prima facie* the most appropriate approach to site waste minimization plan development. The author leans towards this approach during the course of the survey results being assessed, and this contention will need to be evaluated by further literature review, but there is a pattern emerging of respondent comments that suggests this approach

(continued)

Table 7.48 (continued)

No.	Respondent opinions on possible waste minimization strategies	Key points from the author's reflexivity journal whilst developing codes/themes/sub-themes
Q34	Provide other examples of situations where methods, processes or ideas were implemented on your construction site projects that minimized waste	
Pt2	Re-use of prohibited waste products such as contaminated soil or contaminated refractory bricks within the project boundaries, suitably contained via environmental controls	Very proactive response that not only addresses issue of product re-use but mitigates off-site transport and storage in contaminated landfill environmental issue as well, which is also major cost saving to client and contractor
Pt8	Strong recycling strategies for waste materials	With greatest respect to the respondent for troubling to undertake this survey and with the knowledge that participation in such surveys can be time-consuming, the questionnaire in Part C specifically precluded recycling from what this research addresses under front-end waste minimization. Yet, site waste recycling was the number one option that respondents suggested we should look at, over and above those strategies outlined in the survey. The author needs to keep an open mind on each and every comment coming through from respondents, as all their remarks are very constructive

By P. Rundle, March 2018
[a]NB: Pt. 15 in left column is typical; it refers to the corresponding number of respondent comments for Question 26 in Chap. 6, "Results"

Prior to completing the consolidated summary of the front-end site waste minimization findings, in Sect. 7.3.4, the author has inserted this reflexivity journal to remind him of the necessity to provide permeable data and an unbiased synthesis of respondent survey research data.

Please note that Table 7.7, in Sect. 7.2.4.2, "Key Points from the Author's Reflexivity Journal on Sources of Site Waste from Question 25 Respondent Comments for Thematic Analysis Process" of this textbook, also considers a raft of key points during the initial thematic analysis on the sources of site waste in this book. Many reflections by the author are generally relevant to this thematic analysis of waste minimization strategies.

7.3.5 Consolidated Summary of Front-End Site Waste Minimization Strategy Findings

Section 7.2.1 of this textbook provided a synthesis of research data using the adapted Faulkner and Cabon (1998) site waste identification model as a basis to compare this research with prior academic research on this subject.

The abridged Yates (2013) survey questionnaire used in this book for data research development allowed for a close comparison with the Yates (2013) findings. However, the nature of an evaluation of front-end site waste minimization strategies has meant that this task has not been straight forward, compared to the sources of site waste analysis. Hence, whilst Sect. 7.2, "Major Sources of Site Waste Findings", can be regarded as a process driven view, Sect. 7.3, "Front-End Site Waste Minimization Strategy Findings", is looking at a set of strategies. A strategy in this context can be simply defined as a high-level plan for reaching specific business objectives (Thompson et al. 2014). Accordingly, it is a more complicated exercise to undertake a comparative analysis of strategies, rather than processes. For example, a procurement material requisition process prepared to mitigate potential overordering of project stock material is a site waste source issue, whilst an innovation strategy, used to maximize ICT and to minimize site waste, is a more complicated issue to describe and define. Nevertheless, due to the similarities between USA site waste and Australian site waste, though on a different scale, these site waste minimization strategies have been correlated in this research, which enabled a comparative analysis.

What these findings have shown in this section on front-end site waste minimization on Australian projects can be summarized as follows:

I. The "management" theme (92 comments), followed by the "design" (49 comments) and "procurement" themes (19 comments), was perceived as the three key themes that had the greatest bearing on successful front-end site waste reduction strategies.

II. The sub-themes are patterns discernible within the primary themes. The most critical sub-theme was the "FEED" sub-theme, with 45 attributable respondent comments; several of the field supervision management cadre signified the importance of early design initiatives as the key to front-end site waste minimization. The next critical sub-theme, "innovation", with 34 attributable comments covering project management ICT support; technical solutions; formwork solutions; and packaging solutions, was seen by several of the project management and engineering technical cadre as important to site waste minimization. The "client responsibility" sub-theme, with 33 respondent remarks, was the third most critical sub-theme, virtually level with the above-noted "innovation" sub-theme. There was a perception among participants that the client needed to lead the way on site waste reduction. With 27 respondent comments concerning strategies to "maximize off-site construction fabrication and pre-assembly", this theme was viewed by participants as a powerful strategy to minimize waste at the front-end.

III. Fifty per cent of respondents (who wrote remarks for Question 34 requesting other potential front-end waste reduction strategies) could not produce any new waste minimization strategies, further to those noted in Question 26 to Question 33, inclusive, in the survey questionnaire they completed. Fifty per cent of the other respondents said there were no more strategies or were unaware of other strategies.

The following front-end site waste minimization strategies are summarized, in descending order, ranked by the respondent's survey ratings ("yes"/"no"/"don't know"):

   I. Address waste generation reduction during project pre-planning to utilize designs that minimize waste using any of the following techniques: precast; prefabrication; pre-assembly or modularization.

  II. Regarding the use of temporary construction materials, consider waste minimization processes. For example, for concrete construction, designers should specify concrete elements of similar dimensions where practical; use steel shutters on repetitive formwork; ensure that formwork is adequately treated and robustly fabricated, to allow re-use; and confirm that orders are "just in time" to reduce material losses on site.

 III. Adopt a structured approach both for engineering design and in the determination of construction methodologies that involve waste minimization strategies.

 IV. Innovative designs, construction components, or construction processes, must be integrated into construction projects to reduce site generated waste.

  V. The firm must use techniques that improve resource efficiency, equipment efficiency and material resource efficiency and allow for training of manual labour.

 VI. The contractors/builders, consultants and vendors constructively work with the client to minimize change orders that make pre-ordered products redundant and suitable only for waste.

VII. Builders, contractors and engineering consultants should ensure a minimum amount of permanent and temporary materials are expended in the effective provision of client conforming construction/building products.

VIII. The contractor/builder and/or client should have a mandatory waste minimization plan developed as part of the project execution plan.

Section 7.3.3.1, "Summary of Findings for Analysis of Respondent Responses on Front-End Waste Minimization Strategies", considered "*the contractor/builder and/or client having a mandatory waste minimization plan developed as part of the project execution plan*" as only the seventh most likely front-end waste minimization strategy in the Australian engineering construction industry; effectively, it is the least most likely strategy. According to their research on the Australian engineering construction industry, Udawatta et al. (2015) state that a good company waste management plan for project use is considered only the tenth most important strategy in their research on site waste. Back in 1998, Faniran and Caban's (1998) landmark study on Australian project sources of site waste indicated that 57.1 per cent of companies polled had a waste management plan.

A recent paper by Yates (2013) on USA project waste yielded information indicating that 50 per cent of USA's major contractors and commercial builders had a waste management plan in place. Notwithstanding this, this textbook has shown that the Australian engineering construction industry places the highest value on pre-planning that includes engineering design to maximize waste reduction. Yates' (2013) research posits that USA's major constructors put the highest value on standardizing designs and construction methodologies with a view to reducing waste. Further, Udawatta et al. (2015) argue that design practices, with a view to reducing site waste, are the fourth most important strategy.

Telephone calls were made to several of the survey participants in response to the respondent survey answers concerning the need for an innovative site waste minimization strategy to reduce packaging and pallet waste related to the delivery of permanent and non-permanent construction materials. It is evident that survey respondents also rated packaging and pallet waste as the third most important source of site waste (refer to Table 7.2, "Ranking of Potential Sources of Construction Waste by Scoring Each Response"). This led the author to the conclusion that further discussion and research were required in this area to draw a proper research conclusion on packaging and pallet waste reduction strategies.

In the following Chap. 8, "Discussion", the author shall:

I.   Rerank these site waste minimization strategies based on the academic and professional literature to assist the author in their further analysis.
II. Provide interfaces for discussion between themes and sub-themes.
III. Consider the viability of obtaining extra waste minimization strategy survey data on whether there were any packaging waste strategies being adopted by the respondents on their projects, thereby addressing the respondent comments and subsequent telephone conversations with the survey participants regarding this type of waste.

References

Billiton, B. H. P. (2007). *Project manual for capital investment projects*. Australia: BHP Billiton Project Management Centre.

Blayse, A., & Manley, K. (2004). Key influences on construction innovation. *Construction Innovation, 4*(3), 143–154.

Bossink, B. A. G., & Brouwers, H. J. H. (1996). Construction waste—quantification and source evaluation. *Journal of Construction Engineering and Management, 122*(1), 55–60.

Cavalante, S. A. (2013). Understanding the impact of technology on a firm's business model. *European Journal of Innovation Management, 16*(3), 283–300.

Chandler, D. (2016, October 12). Construction waste is big business. *Construction News*, viewed May 2017–May 2018, https://sourceable.net/construction-waste-is-big-business/.

Chignell, B. (2017). *Seven reasons why your employees lack of productivity is your fault*, viewed May 2017, https://www.ciphr.com/advice/7-reasons-employees-unproductive-blame/.

Creswell, J. W., & Plano Clark, V. L. (Eds.). (2011). *Designing and conducting mixed methods research*. Thousand Oaks, USA: Sage.

Creswell, J. W., Plano Clark, V. L., Guttman, M., & Hanson, W. (2003). Advanced mixed methods research designs. In A. Tashakkori & C. Teddlie (Eds.), *Handbook of mixed methods in social & behavioral research* (pp. 209–240). USA: Thousand Oaks, Sage.

Creswell, J. W., & Poth, C. N. (Eds.). (2017). *Qualitative inquiry and research design—Choosing among five approaches*. USA: Sage Publications.

Davidson, C. (2013). Innovation in construction—Before the curtain goes up. *Construction Innovation, 13*(4), 344–351.

Dibner, D. R., & Lerner, A. C. (1992). *Role of public agencies in fostering new technology innovation building*. USA: National Academy of Sciences Press.

Doab, M. (2018). Average cost of construction site waste on major Australian projects. *Doab Estimation Enterprises Proprietary* D'Est Estimating Model.

Egbu, C. (2004). Managing knowledge and intellectual capital for improved organizational innovations in the construction industry: An examination of critical success factors. *Engineering, Construction and Architectural Management, 11*(5), 301–315.

Faniran, O. O., & Caban, G. (1998). Minimizing waste on project construction sites. *Engineering Construction and Architectural Management, 17*(1), 57–72.

Finfgeld-Connett, D. (2010). Generalizability and transferability of meta-synthesis research findings. *Journal of Advanced Nursing, 66,* 246–254.

Gann, D. M. (2000). *Building innovation: Complex constructs in a changing world*. London, UK: Thomas Telford.

Gavilan, R. M., & Bernold, L. E. (1994). Source evaluation of solid waste in building construction. *Journal of Engineering and Management, 120*(3), 536–555.

Gibb, A. G. F. (1999). *Off-site fabrication; pre-fabrication; pre-assembly & modularization*. UK: James Wiley & Sons Inc.

Green, A., & May, S. (2003). Re-engineering construction—Going against the grain. *Building Research and Information Journal, 31*(2), 97–106.

Guest, G. (2012). *Applied thematic analysis*. Sage, USA: Thousand Oaks.

Hampson, D., Kraatz, J. A., & Sanchez, A. X. (2014). *R&D investment and impact in the global construction industry*. Oxon: Routledge.

Heikkinen, D. (2017). *Bad management practices that cause employee inefficiency*, viewed February 2018, https://www.iris.xyz/leadership/bad-management-practices-cause-employee-inefficiency.

Hickey, J. (2015). *Landfilling construction waste in Australia*, viewed April to December 2017. https://sourceable.net/landfilling-construction-waste-in-australia/.

Igo, T., & Skitmore, M. (2006). Diagnosing the organizational culture of an Australian engineering consultancy using the competing values framework. *Construction Innovation, 6*(2), 121–139.

Jaffar, N., Abdul Tharim, A., & Shuib, M. N. (2011). Factors of conflict in the construction industry—A literature review. *Procedia Engineering, 20,* 193–202.

Jaillon, L., Poon, C., & Chiang, Y. (2009). Quantifying the waste reduction potential of using prefabrication in building construction in Hong Kong. *Waste Management, 29,* 309–320.

Jiamei, R. (2012). *Lean construction supply chain*. Denmark: Royal Institute of Technology.

Khalfan, M., McDermott, P., & Cooper, R. (2004). Integrating the supply chain within the construction industry. *Association of Researchers within the Construction Industry, 12,* 897–904.

Langston, C. (2012). *Comparing international construction performance*. Australia: Institute of Sustainable Development & Architecture, Bond University Publishing.

Larson, E., & Gray, C. (Eds.). (2011). *Project management: The management process*. New York: McGraw-Hill/Irwin, The McGraw-Hill Companies Inc.

Lewin, K. (1947). *Field theory in social science*. London: Social Science Paperbacks.

Miller, M., Furneaux, C., Davis, P., Love, P., & O'Donnell, A. (2009). *Built environment procurement practice: Impediments to innovation and opportunities*. Built Environment Industry Innovation Council Report, viewed April 2017, https://eprints.qut.edu.au/27114/1/Furneaux_-_BEIIC_Procurement_Report.pdf.

Morgan, D. L. (2008). *The SAGE encyclopedia of qualitative research methods*. USA: SAGE Publications Inc.

Murphy, M., Perera, R., & Heaney, S. (2008). Building design innovation. *Journal of Engineering, Design and Technology, 6*(2), 99–111.

Myers, S., & Marquis, D. (1969). *Successful industrial innovations: A study of factors underlying innovation in selected firms*. Washington, DC. USA: Government Printing Office.

Nagapan, S., Rahman, I., & Asmi, A. (2012). Factors contributing to physical and non-physical waste generation in construction. *International Journal of Advances in Applied Sciences, 1*(1), 1 10.

National Research University—Higher School of Economics (NRUHSE). (2013). *Innovative construction materials and technologies: their influence on the development of urban planning and urban environment*. World experience—Research Report, The Russian Federation, NRUHSE, Moscow.

Palinkas, L., Horwitz, S., Green, C., Wisdom, J., Naihua Duan, M., & Hoagwood, K. (2015). Purposeful sampling for qualitative data collection and analysis in mixed methods implementation research. *Administration Policy Mental Health, 4*(5), 533–544.

Patton, M. (2005). *Qualitative research findings*. USA: Wiley.

Peer Review Meeting Two. (2018, July 17). *Short list peer review meeting to determine go forward case minutes*. SCU Research File (Peter G. Rundle MSc (Engrg Rsch).

Ritchie, M. (2016, April 20). State of waste 2016—Current and future Australian trends. *MRA Consulting Newsletter*, viewed March to April 2018, https://blog.mraconsulting.com.au/2016/04/20/state-of-waste-2016-current-and-future-australian-trends/.

Ritchie, M. (2017). State of waste data. *MRA Consulting Newsletter*, viewed March to April 2018, https://blog.mraconsulting.com.au/2017/03/29/the-state-of-the-waste-data/.

Rockwell. (2015). *Front-end engineering & design*. USA: Rockwell Automation, USA.

Saldana, J. (2009). *The coding manual for qualitative researchers*. Thousand Oaks: Sage Publications, USA.

Sayfullina, F. (2010). Economic and management aspects of efficiency rise of innovative activity at building enterprises. *Creative Economy, 10*(46), 87–91.

Slaughter, E. (1993). Innovation and learning during implementation: A comparison of user and manufacturer innovations. *Research Policy, 22*(1), 81–95.

Slaughter, E. (1998). Models of construction innovation. *Journal of Construction Engineering and Management, 124*(3), 226–231.

Spivey, D. (1974). Construction solid waste. *Journal of the American Society of Civil Engineers, Construction Division, 100,* 501–506.

Sugiharto, A., Hampson, K., & Sherid, M. (2002). *Non value adding activities in Australian construction projects*. Paper presented to the International Conference for Advancement in Design, Construction, Construction Management and Maintenance of Building Products, Griffith University, Australia.

Sundqvist, E., Backlund, F., & Chroneer, D. (2014). What is project efficiency and effectiveness? *Procedia—Social and Behavioral Sciences, 119,* 278–287.

Suprun, E., & Stewart, R. (2015). Construction innovation diffusion in the Russian Federation. Barriers, drivers and coping strategies. *Construction Innovation, 15*(3), 278–312.

Tatum, C. B. (1991). Incentives for technological innovation in construction. In L. M. Chang (Ed.), *Preparing for Construction in the 21st Century 1991, Proceedings of the Construction Conference, ASCE* (pp. 447–452). New York, USA.

Thompson, A., Peteraf, M., Gamble, E., & Strickland, A. J., III. (2014). *Crafting & executing strategy*. New York: McGraw Hill/Irwin.

Udawatta, A., Zuo, J., Chiveralls, K., & Zillante, G. (2015). Improving waste management in construction projects: An Australian study. *Resources, Conservation and Recycling, 101,* 73–83.

Yates, J. (2013). Sustainable methods for waste minimisation in construction. *Construction Innovation, 13*(3), 281–301.

Chapter 8
Discussion

8.1 Survey Saturation and Survey Sample Size Discussion

8.1.1 Survey Saturation Discussion

Section 6.3, "Survey Saturation", provided compelling information supporting the fact that survey saturation has been met under any academic criterion for this requirement. Charmaz (2006) looked for 25 responses from prospective survey participants. This research survey was issued to 102 participants, of whom 48 respondents was the lowest response rate in the Part 3 survey questionnaire. The author distributed 102 surveys to possible participants aiming for a possible/probable 25 per cent response, but the end result comprised a very high 47 per cent response rate. This nearly doubles the Charmaz (2006) hurdle number of 25 completed surveys. Creswell (1998) suggests an empirical saturation level of 25 respondents, which again indicates that this research nearly doubles that number for the Part 3 question, with the lowest respondent answer rate of 48 participants of the 102 polled. After a comprehensive review by Mason (2010) of over 500,000 abstracts from Great Britain and Ireland higher research degrees dating back from 1716, the average sample size was 25; again, this research nearly doubled that amount.

Conversely, Morse (1995), in an editorial for the *Journal of Qualitative Health Research*, robustly defends the position that the researcher is obligated to leave no stone unturned, especially during the early study phase, to uncover all possible data related to the research topic, including any outlying information that may be perceived as "insignificant"; richness of data is the ultimate aim of attaining saturation, contrary to attaining an empirically derived number. Morse (1995, p. 147) defines saturation as having "data adequacy and operationalized as collecting data until no new information is obtained", which is in alignment with the seminal saturation approach to sampling as outlined by Glaser and Strauss (1967).

When considering possible waste minimization strategies in this research questionnaire, 46.07 per cent of respondents consistently concurred with the 12 site waste minimization strategies nominated in the Part 2 questionnaire, while only 29.4 per

© Springer Nature Switzerland AG 2019   231
P. G. Rundle et al., *Effective Front-End Strategies to Reduce Waste
on Construction Projects*, https://doi.org/10.1007/978-3-030-12399-4_8

cent of respondents disagreed and the balance replied with simply "don't know". Importantly, when asked for other ways to minimize construction waste, the most common response was "recycling" and there were virtually no other innovations provided by the 24 respondents who listed the possible alternatives to the 12 options covered in the questionnaire. Similarly, for the sources of waste questionnaire in this research the top three most likely sources of site waste and the bottom two least likely sources of site waste were in common with the seminal paper used as a benchmark for this textbook (Faniran and Caban 1998). Again, there were few unique or innovative responses in reply to the question asking respondents to provide information on any other sources of waste than those nominated.

Chapter 6, "Results", of this book clearly demonstrates that this research has provided a rich source of data from respondents. This data was sufficient enough to satisfy both the "empirical" saturation definition (of around 25 survey responses) and the "classic" saturation definition. Such a large pool of data was harvested from respondent questionnaires that a detailed thematic analysis of this information was possible to a level where questions on "other possible sources of site waste" and "other possible site waste minimization" did not provide further possible strategies than those nominated in this research questionnaire; no new potential sources of site waste were identified. The major single other possible waste reduction strategy highlighted by respondents was the end of site material supply chain tried and long-time implemented recycling strategy. However, what the wealth of respondent data did provide in this research, were the keen insights from a broad spectrum of engineering construction industry professionals into current Australian perspectives on addressing these "known unknowns, which was proven in mapping out a path forward to reduce construction site waste at the front-end of the construction materials supply chain.

Even a heuristic view of the respondent sample size, when compared to the benchmark questionnaires this research survey was adapted from, indicates that saturation has been attained. Forty-eight participants responded to the questionnaires for Part 3 of this research, compared to the benchmark Yates (2013) study on site waste minimization strategies, which received 29 survey responses. The Part 2 questionnaire respondent sample pool was 51 respondents versus only 24 completed questionnaires in the Faniran and Caban (1998) benchmark paper on sources of site waste.

Pursuant to the above, by any measure, the qualitative research data harvested on this project has attained saturation.

8.1.2 Survey Sample Size Discussion

Strauss and Corbin (1998) posit that one of the most common questions post-graduate research students pose to their supervisors is "how large should my survey sample be?" The 102 surveys transmitted to potential participants for this research was calculated on a 25 per cent respondent participation rate providing the empirically derived number of 25 surveys necessary to successfully undertake this research (see

Sect. 8.1.1 for derivation of this empirically completed survey size of 25). There were 48 survey questionnaire forms completed for Part 3, 51 survey questionnaire forms finished for Part 2 and 53, Part 1 questionnaire forms completed for this research.

The author notes that a successful return of the commensurate numbers of completed surveys was vital to attaining saturation. Even the Part 3 survey questionnaire (with the lowest response rate) delivered a 47 per cent survey completion rate. This was achieved by sending the questionnaires out individually with a personalized covering email to each potential participant and by following up any participant queries by telephone. Further to this, the return date of the questionnaires was extended, upon request by the participants. Another proactive factor in the large number of participants in this research survey was that the author began contacting potential sample participants five months prior to the release of the pro forma survey questionnaires to respondents. Bearing on the author's rationale for survey sample development was the fact that the seminal Yates (2013) research paper stated that the 150 engineering and construction executives initially contacted to undertake questionnaires, from the association funding this research, all failed to initially provide a single completed survey. The author would adopt the same approach in any further research on this topic using the assumption of a 25 per cent respondent return survey rate. A previous Masters of Business Administration engineering industry research project (that the author conducted) provided only around a 30 per cent response rate.

Accordingly, the approximately 50 completed survey forms, twice as many as were empirically required to attain saturation, provided a rich pool of data to harvest information for both the development of the severity index for sources of site waste and a detailed thematic analysis of the site waste minimization strategies research component. Therefore, the 102 survey forms that were issued to the prospective participants was an appropriate sample size for distribution. This was further validated empirically by Coakes and Ong (2011), who nominated a sample size of 100 or more as being appropriate, albeit for a human factors approach to waste management strategy review, as per the Udawatta et al. (2015) construction waste study.

8.2   Demographics Discussion

The Yates (2013) survey questionnaire, from which Part 3 of this research survey questionnaire was adapted, averaged 46.3 per cent of its 29 respondents answering "yes" to various site waste minimization strategies, while 23.3 per cent of the 29 respondents answered "no" and the balance of respondents indicated "don't know". For this research survey, 46.07 per cent of the 48 respondents to the Part 3 survey averaged "yes", which is in particularly close alignment with the Yates (2013) survey (46.3 per cent). However, 29.4 per cent of the 48 respondents from this research survey answered "no", on average, to the Part 3 questionnaire covering site waste minimization strategies. This is compared with an average of 23.3 per cent of the "no" responses from the total 29 respondent sample of the Yates (2013) survey. The balance of the 48 respondents to this Part 3 questionnaire who answered neither "yes" nor "no", responded "don't know".

Section 7.2.2, "Comparison of Part 2 Survey Data with Other Academic Journals for Sources of Site Waste", notes that the top three (most likely) sources of site waste and the bottom two (least likely) sources of site waste for the 24 respondents polled in the Faniran and Caban (1998) survey on sources of site waste, concurs with the survey results for this research, which received 51 responses for this surveys Part 2 questionnaire.

There was a possible risk with the research approach adopted by the author to mitigate sampling bias by utilizing a stratified sampling methodology. In this methodology, members of the sample population who had previous experience in the engineering construction industry were divided and formed into homogeneous sub-groups (Kish 1995). These participant sub-groups, comprising both various organization positions and sectors of work within the engineering construction industry, could possibly have undertaken the survey along these cohort lines. As an example, participant roles varied as follows:

I.   Company directors; CEOs; Chief Operating Officers; a chairperson; and group general managers;
II. Project directors;
III. Project managers;
IV. Site or construction managers;
V.   General superintendents;
VI. Engineering design managers/lead architects;
VII. Technical managers (a quantity surveyor; an estimator planner/scheduler); and
VIII. "Others" (managers for sustainability/environmental/risk/Quality Assurance (QA) and contracts, plus general fore-people and academics with an engineering background).

These respondents have also come from various fields within the engineering construction industry, as follows:

I.   Heavy engineering construction including infrastructure—24 respondents;
II.   Commercial and residential building, including high-rise apartments;
III. Mining and resources capital works;
IV.   Project sustainability;
V.   Private sector clients;
VI.   Public sector clients;
VII. Public sector waste management personnel;
VIII. Engineer academics with field experience; and
IX.   Engineer academics with site waste management experience.

The above-mentioned risk, in using such a broad demographic sample of engineering construction personnel, is that individual allegiances would be split into particular respondent fields of expertise. However, this has proven not to be the case, as can be seen in this Sect. 7.3. Rather remarkably, 46 per cent of the baseline Yates

(2013) survey participants and 46 per cent of the respondents for this research on the topic of site waste minimization strategies agree on their merit, whereas the top three (most likely) sources of site waste and the bottom two (least likely) sources of site waste for the 24 respondents polled in the Faniran and Caban (1998) survey on sources of site waste, concur with the survey results for this textbook.

The benefit of using a stratified sampling strategy divided into homogenous subgroups has not only reduced a propensity for sampling bias, but also positively benefited this research because the respondents provided a broad spectrum of opinions from their own fields of the engineering construction industry. This has provided significant insights into the broadly agreed upon sources of site waste and site waste mitigation strategies.

8.3 Research Question Discussion

8.3.1 Revised Research Question

Section 3.4.2 provides the development of the revised research question for the "Go Forward" case that was ultimately selected from the eight option 'long list' after a process of knowledge acquisition via a literature review and an independent peer review assessment using the Kepner Tregoe decision analysis formal process. The revised research question was: *"what effective and efficient front-end strategies are available to reduce waste on Australian construction projects?"* This discussion focuses on validating that this research has satisfactorily answered the revised research question.

The definition of effectiveness used in this textbook is: "satisfying or exceeding customer needs with a compliant product" (Sundqvist et al. 2014, p. 281). Efficiency is defined as: "meeting all internal requirements for cost, margins, asset utilization and other related efficiency measures" (Sundqvist et al. 2014, p. 281). Refer to Sect. 1.2, "Definition of Effectiveness and Efficiency", for more information.

The reader's attention is drawn to Sect. 7.3.5, "Consolidated Summary of Front-End Site Waste Minimization Strategies Findings", which indicates that the 48 respondents could add no new site waste reduction strategies to those strategies denoted in the abridged Yates (2013) front-end site waste minimization strategies model. When asked in survey Question 32 to provide any other front-end construction waste strategies, respondent comments provided versions of those strategies already noted on the survey questionnaire. Several respondents highlighted "recycling", which is a "back-end" site waste material logistics chain activity and, hence, specifically excluded in the survey scope for this textbook; however, this is only part of the answer.

Section 8.6, "Front-End Site Waste Minimization Strategies Discussion", provides a detailed appraisal of the raft of research results obtained from the analysis of the model data from 48 respondents using thematic analysis, and importantly, places some context around this discussion underpinned by the literature. The "Conclusions", Chap. 9, and "Recommendations", Chap. 10, capture the path forward for front-end site waste minimization on Australian projects uncovered in this research, with suggestions for further research on this topic.

The strategies have been adequately outlined in the research. However, it is important to ascertain if they are effective and efficient. The definition of an effective strategy is one that meets or exceeds customer needs in the provision of a compliant product. The research has shown that savings of around two per cent of the contract price may be enjoyed by the implementation of these site waste reduction strategies. The construction waste minimization strategies recommended in this research are also effective because they involve using pre-existing available design and construction practices that are not reliant on any new and high-risk methodologies. In today's business world, corporate social responsibility has an enormous impact on share prices (Grossman 2005).

The adoption of front-end waste minimization practices, as highlighted throughout this textbook, may have many ecological and social benefits, including but not limited to, a reduction in carbon emissions due to a decrease in landfill footprints (Chandler 2016; Hickey 2015), and, as the author argues, a potential decrease in public housing construction costs. The estimated two per cent savings in construction costs may provide significant commercial benefits to key stakeholders, further validating the effectiveness of these waste reduction strategies (Vowles 2011).

These front-end site waste minimization strategies will also need to be efficient practices that shall meet internal requirements for cost, margins, asset utilization and other related efficiency measures. The construction waste reduction strategies, outlined in the Chap. 9, "Conclusions", and Chap. 10, "Recommendations", provide a number of author proposals that could measurably improve project efficiencies that could have the potential to increase contract savings by even more than the mooted two per cent. The implementation of a project procurement plan to reduce material waste, by providing procedures to minimize ordering and take-off errors and developing an integrated project material logistics chain, are all efficiency improvements. The recommendation is that projects use a design procedure that directs that the designers involve the contractors early on in the engineering development to capture efficient construction methodologies, such as the use of precast concrete elements, and design to suit.

It is also important to note that the low respondent "yes" response in this research survey, when the Australian participants were asked if there was a focus on improving company efficiency in equipment, labour and materials use, compared to a USA "sister" study, undertaken using the same model (Yates 2013), alerted the author to perform a further literature review. This review clearly showed that Australia is lagging behind the USA in construction efficiency (Langston 2014).

This research found that the major source of construction waste was believed to be design changes initiated by the client. In the front-end waste research survey, participants were requested to remark on the optimum method to reduce client scope changes and the most common sub-theme was "communication". Hoezen et al. (2006, n.p.) believe that "[t]he efficiency and effectiveness of the construction process strongly depend[s] on the quality of communication". The importance of the inclusion of a communications plan, for all capital works projects, has been made in Chap. 9, "Conclusions".

Pursuant to the above, the revised research question has been answered satisfactorily and in considerable detail.

8.3.2 Research Sub-question

According to the qualitative methods pragmatic framework approach, it was necessary to prepare a research sub-question supplementary to the (revised) research question, which was developed to focus on addressing the research on physical site waste sources and is complementary to the research on construction waste minimization. The Swinburne University (2016) engineering faculty approach to research question development was used to develop the research sub-question, which asks: "*what are the major sources of waste on Australian construction sites?*"

Section 7.2.3, "Summary Findings on Part 2 Question 12 to 24, Sources of Site Waste", summarizes the research findings from the respondent survey answers to Likert Scale Questions 12 to 24, inclusive, which evaluated 13 potential site waste sources on Australian construction projects based on the Faniran and Caban (1998) survey questionnaire used for this research. The respondents were asked to evaluate the 13 potential sources of waste as being either (i) "not significant"; (ii) "of minor significance"; (iii) "significant"; or (iv) "very significant".

The evaluation synthesis of the available research data indicates that the top six causes of waste are denoted as:

    I. Client initiated changes;
   II. Design and detailing errors;
  III. Packaging and pallet waste;
  IV. Procurement ordering and take-off errors;
   V. Lack of on-site materials planning and control; and
  VI. Improper material storage.

Pursuant to the above, the research sub-question has been effectively answered in this research.

8.4   Pragmatic Qualitative Research Framework Discussion

8.4.1   Summary Table of Pragmatic Qualitative Framework Discussion

Table 8.1, "Framework for a Pragmatic Research Approach Discussion", adapted from Pope et al. (2000) from British research standards, clearly summarizes that the research framework has provided results via rigorous academic research methodologies stringently following the nominated qualitative research approach.

8.4.2   Framing Discussion

The framework for the pragmatic approach for this textbook states that framing must provide clarity of what is known and of what this study builds upon (refer to Table 8.1). Refer to Sect. 2.2.5, "Australian Construction Site Waste", Sect. 3.2.2, "Option B 'Short-List'—Construction Site Waste Reduction", and Sect. 5.1.1, "Site Waste Sources and Minimization", for the full details of what this study builds upon.

This book has provided a number of unique research aspects from both an Australian and an international context, while adopting a framework that has made minor adaptations to two existing models and allowing a comparison with the existing data. In addition, this research evaluates data on the sources of site waste and examines possible construction waste reduction strategies—both of these subjects are generally addressed in the current research.

Recommendations have been made for minor configuration changes to the pre-existing models to make the data output more current. A thematic analysis approach was used to interpret a rich pool of data from the respondents, rather than by simply collating this information in tabular format. Finally, no other paper on construction waste reduction/sources of site waste cited in the "References" section of this book has had such a broad spectrum of research survey participants to enable opinions from across the engineering construction industry, both in Australia and/or overseas.

This research has shown that the three major sources of site waste, as well as two of the least likely sources of site waste of the 13 potential sources, researched by Faniran and Caban in their 1998 Australian paper, are the same sources of waste rankings as this research. Further, there is wide consensus in the Australia engineering construction industry regarding the most appropriate strategies to reduce waste, as per the recent USA academic research within large contracting and non-residential building sectors. This situation is alarming because construction and demolition waste is the major generator of waste in Australia and a reduction at the front-end could result in significant savings to the engineering construction industry and to the client stakeholders bottom line (Chandler 2016). Importantly, it could be of major social and environmental benefit, by reducing the landfill footprint and carbon emissions, manufacturing/transportation energy and water consumption (Chandler

Table 8.1 Framework for a pragmatic qualitative research approach discussion

Framework for a pragmatic approach

Standards	Terms associated with qualitative research	Function	Remarks	Discussion
Framing	–	Show the research has purpose in the context	Make it clear when "framing", what is already known and what this study builds on. Also, what is the "something" that has not been described before	Refer to Sect. 8.4.2, "Framing Discussion", which provides details on what is known and what this study builds on, along with the information never described before
Reliability	Procedural trustworthiness	Ensure respect for procedures	Justify the size and nature of the sample. The data must properly address the research question	Refer to Sects. 8.1 and 8.2, which justify the size and nature of the sample. Refer to Sect. 8.3 which addresses the research question
Validity	Trustworthiness of the findings	Ensure respect for findings	Ensure analysis is realistic—use techniques associated with grounded theory (i.e. thematic analysis using open coding etc.). Ensure findings are coherent with categories defined and mutually exclusive. Do the findings show theoretical validity by developing theory? Do findings show catalytic validity by demonstrating capability to change participants?	Refer to Sect. 8.4.3, "Validity Discussion", which details the triangulation of data from academic literature; previous academic research was used to correlate this data. This sub-section also details how the sample process has demonstrated catalytic validity

(continued)

Table 8.1 (continued)

Framework for a pragmatic approach

Standards	Terms associated with qualitative research	Function	Remarks	Discussion
Generalisability	Transferability	Results are useful to other practitioners	Will new concepts be transferable? Is research for practitioners and/or academics?	Refer to Sect. 8.4.4, "Generalisability Discussion", which details exactly how seminal international models adopted for this research have allowed for transferability. This sub-section also details how KPIs of project success is the practical application of this research for use in the field for practitioners as well as academics.
Objectivity	Permeability/trustworthiness of researcher	Reflexivity and protection from unwanted bias	Will researcher be open to change provided by the research and will the researcher rigorously test the analysis?	Refer to Sect. 8.4.5, "Objectivity Discussion", for full details. During 40 years as a civil engineer, the author has worked on over 20 feasibility studies and this has taught him not to assume a result but to allow the study to provide the options. A reflexivity journal was kept and the details are contained in Tables 7.7 and 7.40 of this book containing the author's key reflections on respondent comments covering the questionnaire on sources of site waste and the questionnaire on site waste minimization strategies. The author has also taken self-awareness training as part of a Masters of Business Administration and has used these practices on reflecting on this textbook's approaches. An independent facilitator and two independent peer review panels containing eminent specialists were adopted in this research. Research findings were rigorously tested against pre-existing academic research with close correlations being determined

By: P. Rundle, 2018 (adapted from: Pope et al. 2000)

2016; Hickey 2015). Chapter 10, "Recommendations", of this book maps out a path forward for implementing site waste reduction changes.

8.4.3 Validity Discussion

The research analysis has proven to be realistic, with thematic analysis commencing with open coding of respondent comments being a qualitative methodology associated with grounded theory. This process involved the critical review of the responses to determine appropriate open coding used for the purposes of the formation of themes and sub-themes from these codes (Braun and Clark 2006).

Respondent comments developed for this research correspond to a code, which has been bracketed into a theme that was largely selected from a table developed by Nagapan et al. (2012, p. 5, Table 2.2). This was possible because Nagapan et al. (2012) followed the seminal approach established by Spivey (1974) on site waste and developed by later academic papers, whilst developing codes and corresponding primary themes and supported by an extensive global literature review of site waste. The sub-themes were synthesized by the author and, where the Nagapan et al. (2012) themes were deemed inappropriate, the author carried out their own theme determination, similarly for any new themes not on the Nagapan et al. (2012) Table 2.2. Accordingly, findings are coherent with categories that are separate and different from each other.

The Faniran and Caban (1998) model (for the identification of potential sources of site waste) was used in this research to determine the current situation on Australian construction projects regarding possible sources of site waste. This model data was then used in the interpretation of respondent remarks using thematic analysis and the Faniran and Caban (1998) recommended severity index theory. Similarly, an adapted Yates model (2013) was used in this research to determine possible front-end site waste minimization strategies that could be utilized on Australian projects, with this model data then used in the interpretation of respondent remarks using thematic analysis. The advantage of this approach was that, coupled with a detailed literature review of available academia, professional and trade literature on site waste, and pre-existing research data from previous academic papers, (Faniran and Caban 1998; Nagapan et al. 2012; Yates 2013) these research findings could be regularly benchmarked and triangulated against other studies, thus considerably aiding in the validity of this textbook.

Forty-six per cent of the baseline Yates (2013) survey participants and 46 per cent of the respondents for this research on the topic of site waste minimization strategies agree on their merit. The top three (most likely) sources of site waste and the bottom two (least likely) sources of site waste (for the 24 respondents polled in the Faniran and Caban (1998) survey on sources of site waste) concur with the survey results for this research. The benefit of using a stratified sampling strategy divided into homogenous sub-groups has not only reduced a propensity for sampling bias, but also positively benefited this research since the respondents provided a broad spectrum of opinions from their own fields of the engineering construction industry, therefore, providing significant insight into the broadly agreed sources of site waste and site waste mitigation strategies.

These findings show that this research has catalytic validity that demonstrates the capability to change participants within the broad spectrum of the Australian engineering construction industry.

8.4.4   Generalisability Discussion

The research undertaken for this textbook is transferable to future research both within Australia and internationally. The reason for this is that the Faniran and Caban (1998) Australian site waste source identification model was developed from Spivey's (1974) seminal work on construction waste classification and built upon by Gavilan and Bernold (1994) in the USA. Nagapan et al. (2012) used an adapted form of the Faniran and Caban (1998) site waste identification model, and hence, the author has been able to readily compare his research results with past research data. Naturally, the economic and climatic environmental conditions must be factored into any assessment of site waste using the above noted model. The reader's attention is drawn to Sect. 7.2.2, "Comparison of Part 2 Survey Data with Other Academic Journals for Sources of Site Waste", of this book, which takes a necessarily circumspect view of comparing the Nagapan et al. (2012) Malaysian model results for sources of site waste with the author's 2017 Australian survey data using virtually the same model and the differences are explained by possible economic and climatic variables. Reference is made to Sect. 8.2, "Demographics", in which it has been plainly demonstrated that there is a nexus between the results from this research and the adapted Yates (2013) seminal USA model results on construction waste minimization, as well as other academic papers (Faniran and Caban 1998; Nagapan et al. 2012; Sugiharto et al. 2002) covering site waste sources and/or construction waste reduction.

Pursuant to the above discussion, it can be seen that there is a strong element of transferability available for use in future research and, to look back at past research into construction site waste, both within Australia and internationally, that due prudence has been used for economic and climatic variables between the research locations being evaluated.

It would be valuable to review Sect. 1.4, "Key Performance Indicators", of this book to address matters of generalisability and more particularly, the value of this textbook to both the academic and professional worlds. These KPIs are noted as follows: (i) the research will be of transparent commercial benefit to contractor/engineering houses, client end users and the community at large; (ii) the proffered solutions can be readily and therefore, expediently implemented; (iii) the topics selected for investigation shall provide maximum stakeholder benefits; (iv) the solutions to the identified inefficiencies and ineffective practices are, by and large, available within the academic and professional international body of knowledge;

(v) the textbook be must be practical in nature and address a void in the Australian engineering construction business and the work must be valuable to this industry; and (vi) the identified research must broadly comply with a triple bottom line philosophy (Slaper and Hall 2011) and commercial, social and ecological benefits must be variously enhanced, as appropriate, via these initiatives.

In accordance with the above, it can be seen from the minimum KPIs, upon which the eight original proposed construction methodologies were evaluated for eventual further research of one option, front-end site waste reduction, the relevance of this research to the Australian engineering construction industry was a key criterion.

8.4.5   Objectivity Discussion

Stiles (1993, pp. 607–611), writing on quality control in qualitative research, noted that qualitative investigators explicitly reject the possibility of absolute objectivity and truth. The concept of objectivity is replaced by the concept of permeability—the capacity of understanding to be changed by encounters with observations (Stiles 1993). Investigators argue that we cannot view reality from outside of our own frame of reference (Stiles 1993). Instead, good practice in research seeks to ensure that understanding is permeated by observation (Stiles 1993). The validity of an interpretation is always in relation to some person, and the criteria for assessing validity depends on who that person is (for example, the reader, investigator or research participant) (Stiles 1993, pp. 607–611).

The author has managed over 20 major engineering and mining feasibility studies, the largest study costing US$58 million to execute. During the course of carrying out these studies, the author learned that one cannot pre-suppose a study's commercial and technical outcomes. It is not possible to commence at the end result and work backwards; hence, the scoping study, concept study, prefeasibility and feasibility staged approach to project development are important. This attitude was brought to bear when developing this research. The results have told the story, the author had no preconceived perception on outcomes until the research findings analysed what the "picture" was and, in fact, was amazed on occasion, even at the literature review stage, concerning several of the outcomes.

The author employed many practices to mitigate self-bias that could have impacted permeability because of his 40 years of experience as a civil engineer. Some of the mechanisms employed were:

I.   The use of an independent facilitator and an independent panel of peer reviewers from different sectors of the engineering construction industry to select a four option 'short-list', from an eight option 'long-list' developed from a literature review brief by the author using a risk-based decision approach.

II. Reference to a separate independent panel of eminent peers from the engineering construction industry and engineering academia to select a "Go Forward" case from a four option 'short-list', using detailed literature review briefs prepared by the author. A Kepner Tregoe risk-based decision-making approach was undertaken using an independent facilitator via a virtual skype workshop of 10 attendees.

The author utilized prior self-awareness training to incorporate a reflexivity journal as part of the thematic analysis component of this research. Refer to Sect. 7.2.4.1, "Key Points from the Author's Reflexivity Journal on Sources of Site Waste from Question 25 Respondent Comments for Thematic Analysis Process", for an example of the key points from this log used for the study of potential sources of site waste.

Another example of the author's mitigation of sample bias is detailed in Sect. 8.2, "Demographics Discussion". This sub-section describes, in depth, how a broad spectrum of engineering construction industry personnel were canvased as respondents for this research survey.

Section 7.2.2, "Comparison of Part 2 Survey Data with Other Academic Journals for Sources of Site Waste", Sect. 8.1, "Survey Saturation and Survey Sample Size Discussion", and Sect. 8.2, "Demographics Discussion", all provide examples of the researcher rigorously testing the analysis using previous benchmark research.

8.5   Sources of Site Waste Discussion

The following six main options have been noted as the major sources of construction site waste from an analysis of survey data for this research obtained via an analysis of the respondent completed survey questionnaire, Part 2, "Sources of Site Waste".

  I.   Client initiated design changes;
 II.   Design and detail errors;
III.   Packaging and pallet waste;
 IV.   Procurement ordering and take-off errors;
  V.   Lack of on-site materials planning and control; and
 VI.   Improper material storage.

Table 6.3 in Sect. 6.5.3, "Respondent Comments Covering Other Sources of Waste", provides respondent comments to the final question, Question 25 of Part 2 of the research survey, which requests respondents to "*list any other sources in your opinion that cause site construction waste*". A frank evaluation of these respondent comments reveals that there were no new sources of waste other than the 13 potential sources nominated in the adapted Faniran and Caban (1998) site waste identification model used in this research survey. However, a thematic analysis was undertaken on these Question 25 respondent comments and, as an exercise, an adapted thematic analysis was carried out on the list of the top six causes of site waste determined via a severity index analysis, along with an evaluation of respondent answers to a Likert Scale test, based on Faniran and Caban's (1998) site waste identification model. The thematic analysis derived themes that this research points to as being the most relevant to site waste sources, in descending order, as follows:

   I. Management;
  II. Procurement;
 III. Design; and
  IV. Workers.

Research by thematic analysis on sources of site waste found that the four sub-themes involved in site waste sources, in descending order, are as follows.

   I. Quality process shortfall;
  II. Front-end engineering design (FEED);
 III. Work culture; and
  IV. Client responsibility, which rated equal third with the "work culture" sub-theme.

The author argues that there is a strong inter-relationship between themes and sub-themes, that this research has pointed to, over and above the obvious interfaces. The research has shown that the "management" theme is a main control driver for sources of site waste and hence, the primary influencer of change to reduce site waste sources. A secondary outcome of this research indicates that the "client responsibility" sub-theme is considered by the respondents to be an important influence running through this "management" cadre, which has the capacity to change/reduce site waste sources, and by inference, site waste volumes. An especially important component of this change process to reduce sources of waste would be a major shift in top-down engineering construction industry work culture. This could occur through the "design" phase (theme). More particularly, it could be implemented with a concentration on "FEED" (sub-theme) commencing at the feasibility study phase, so that the project building blocks are laid to underpin waste minimization by site waste source control, through to the "workers" (theme) at the construction job face, be they manual, technical or project management "workers".

This research demonstrates that the second key driver of sources of site waste is the "procurement" theme. The issues surrounding procurement (such as attention to handling construction materials to mitigate damage, adequate materials warehousing on site to protect construction materials, defective and/or non-compliant material purchase and reducing ordering/take-off errors) can all be adequately and readily addressed by developing and following project quality assurance processes. Any shortfalls in an inadequately prepared procurement quality plan, or in failing to follow the project procurement plan ("quality process shortfall" sub-theme), will increase site waste source potential and, correspondingly, site waste volume. The solutions are there for this secondary key driver, they just need to be adopted and followed.

As alluded to above, the procurement/quality shortfall may be addressed with limited difficulty via careful planning, good control and the willingness of the client to enter into a contract that has a budget that allows for these procurement quality processes to be followed, along with a contractor/builder who diligently executes these processes rather than into net profit.

The other major theme influencer of sources of site waste is the "workers" theme and the respondents comments relate this main theme to the "work culture" sub-theme. This theme involves the difficult task of altering the focus of the engineering construction industry's work culture to become focused on reducing site waste at the front-end stage of the material logistics chain. This could entail a major cultural shift in Australian project development, engineering design office and construction site work practices by clients, management and workers. Change management on work culture is a difficult process requiring continual reinforcement, as Kurt Lewin's (1947) seminal work eloquently expounds today as it did 70 years ago. However, this industry wide cultural change management process of the Australian engineering construction industry, which represents around 10 per cent of our work force, may be necessary to bring about a significant reduction in site waste volume (DIIS 2017).

In summary, respondent comments on Question 25, regarding other possible sources of construction site waste, did not overtly indicate any new possibilities in excess of the 13 sources in the adapted Faniran and Caban (1998) site waste identification model. However, a synthesis of the data using thematic analysis revealed an in-depth evaluation of the linkages between the identified themes and sub-themes and added to the knowledge acquisition base for this book.

Section 8.5 commenced with a summary of the six major sources of construction waste as determined from an analysis of the data from Questions 12 to 24, inclusive, of the Part 2 survey questionnaire for sources of site waste as developed by the adapted Faniran and Caban (1998) site waste identification model, using the severity index process along with a method for ranking of all respondent data for each question. These six potential sources of site waste are:

   I. Client initiated design changes;
  II. Design and detail errors;
 III. Packaging and pallet waste;
 IV. Procurement ordering and take-off errors;
  V. Lack of on-site materials planning and control; and
 VI. Improper material storage.

Commencing this discussion on the major possible sources of construction waste through the lens of a thematic analysis has enabled the author to focus more cogently on the root causes of these six key areas.

Unambiguously, "client initiated design changes" can be minimized via the correct life cycle planning of a project, commencing at inception, from when a client nominates their business objectives for a project, and with the project scope of works aligning with the client project objectives (Amado et al. 2012). The preliminary phase of a project should comprise of a scoping study to develop the project scoping requirements, a concept study to determine a broad basis on whether a potential project is possibly viable, followed by a pre-feasibility study to select the most feasible option, and finally a feasibility study that provides sufficient engineering design, project cost estimates and a project schedule, along with other deliverables such as environmental, regulatory and financial modelling, to determine the robustness of the proposal. All studies are critical to developing a project that shall meet client objectives and be completed on schedule and on budget (Larson and Gray 2011).

To achieve this process, sufficient time should be allowed during the project development phase, which can take 30 months up to the completion of the preliminary engineering design for a major project over US$500 million (BHP Billiton 2007).

These efforts are not employed solely to mitigate the one source of construction waste in this research and the other papers benchmarked in this research (Faniran and Caban 1998; Nagapan et al. 2012, Sugiharto et al. 2007). Following this structured preliminary process is the key to a positive project outcome, for its entire life cycle into operation, right through to de-commissioning (Larson and Gray 2011).

Naturally, there are projects where the client's views must be constructed on a "fast–track" basis, where political and market forces, for example, can necessitate an accelerated project execution timeframe. An example of this is the Bougainville copper mine project that the Bechtel corporation constructed to ensure that there were royalties and benefits to Australia and to the owners, CRA, before Papua New Guinea obtained independence from Australia in 1975 (Tonks 1981). Another example of a fast-track project was the Goro nickel mine in New Caledonia that commenced construction in 2002. Works were suspended in December 2002 with the engineering, design and procurement joint venture, Bechtel-Technip-Hatch, removed off the project because of budget over runs from US$1.45 billion, from the original contract, up to US$2.0 billion, with a planned construction completion in March 2004 to take advantage of the massive rise in nickel prices used in bullet hardening as a result of wars in Afghanistan and Iraq (Moody 2013).

The Goro project was eventually completed in 2010, for an estimated cost in excess of US$4 billion and, since production commencement, has been dogged with continual delays due to environmental and traditional landowner issues (Moody 2013; Wacaster 2010). Australian contractors and designers were major stakeholders on both the Goro and the Bougainville projects. Anecdotal advice from the Goro client site manager indicated that, when the Goro project re-commenced in 2004, there were major permanent site waste costs absorbed by the project because both the location and design of the process plant was changed, as well as the port design, making all earthworks and equipment orders up to December 2002, redundant, with earthworks alone costing A$70 million by Theiss Contractors.

The following potential sources of site waste, in descending order, fall within an over-arching umbrella of competent project management, including client capital works capability, along with design and procurement quality processes being in place, in the project execution plan (BHP Billiton 2007):

  I. Design and detail errors (No. 2);
 II. Procurement ordering and take-off errors (No. 4);
III. Lack of on-site materials planning and control (No. 5); and
 IV. Improper material storage (No. 6).

There will always be errors in design and design detailing, as well as in equipment and material specifying and ordering. For example, no project procurement plan can guarantee that there will be no accidental off-loading damage to materials. Some materials, such as bagged high strength mortar, always provided in limited supply, will potentially be subject to both storage and handling problems. However, careful attention to detail, following project design and procurement processes, and allowing for reasonable material losses could save millions of dollars on major projects.

The common thread in all five of these potential sources of site waste is that, ultimately, the client has the commensurate funds to buy the necessary expertise. This expertise could be used to properly execute the project and to minimize the potential sources of waste as a beneficial by-product. The author argues that this would result in significant cost savings via a substantial reduction in material waste, as well as social and environmental triple bottom line benefits related to a reduction in landfill foot prints, reduction in carbon emissions and a reduction in residential building costs or a plateauing of those costs.

The "packaging and pallet waste" theme was ranked third, out of the top six possible sources of waste after an evaluation of the respondent completed questionnaire. Packaging and pallet waste are particularly contentious issues in congested work sites such as high rise commercial buildings or apartment building constructions, which has gone through a boom from 2015 to 2017 (ACA 2017). This topic shall be the focus of a recommendation for further study through a peak industry body such as the Australian Construction Industry Forum to see if the member companies could establish a working party to consider the ways in which suppliers could limit packaging without compromising product quality while handling the materials.

8.6   Front-End Site Waste Minimization Strategies Discussion

The consolidated summary of front-end site waste minimization strategies Chap. 7, "Findings", indicated that the surveyed Australian engineering construction participants considered that a mandatory company waste minimization plan was the lowest ranked priority of construction waste reduction strategies adopted by their firm (refer to Sect. 7.3.5, "Consolidated Summary of Front-End Site Waste Minimization Strategies Findings"). Udawatta et al. (2015) state that their Australian survey showed that a company waste minimization plan was considered only tenth, in importance, as a waste minimization strategy. Faniran and Caban's (1998) landmark sources of site waste identification study noted that 57.1 per cent of Australian engineering construction firms interviewed had a company site waste management policy. The Yates (2013) research in the USA posits that 50 per cent of that country's major contractors and commercial builders have a corporate site waste management plan. This research showed that only 37.5 per cent of respondents could answer "yes" and affirm with certainty that their company had a site waste reduction management plan.

Notwithstanding the above, the author posits that a standard company site waste management plan is the most fundamental strategy to reducing construction site waste. The research has shown that engineering construction industry survey participants placed a very high regard on engineering design at the earliest phases as an effective strategy to assist with front-end waste reduction. Udawatta et al. (2015)

notes that their Australian research showed that industry professionals ranked early design fourth in importance for project waste reduction, as did Yates (2013), whose research showed the high emphasis that the USA's major contracting and building firms place on standard designs and methodologies from project inception.

The author suggests that this research is compatible with international literature and has shown that the engineering construction industry presently recognition the value of front-end preliminary design at project inception as an appropriate strategy that assists in reducing site waste. However, a company site waste management plan, which includes all phases of the project process from study stage, through to design, construction execution and operation, must be a mandatory document. The reason for this is that without a standard operating procedure, a project would be largely in the hands of the core competency level of the project manager and design manager on waste mitigation design and construction methodology strategies. This factor, at best, would create a disjointed approach for companies across their projects and at worst, could lead to major failures to prevent large cost over runs due to site waste quantity and scope blow outs. Ameh and Daniel (2013) argue that, on average, material wastage costs construction projects amounts ranging from 21 to 30 per cent in cost over-runs.

Transferring knowledge within the construction sector is historically difficult to achieve because construction projects generate a huge body of knowledge available for sharing that is not readily disseminated amongst peers (Argote et al. 2000). Eisenhardt and Zbaracki (1992) believe that a firm's project portfolio approach is the embodiment of their strategy. Kester et al. (2013, p. 2) note that "to survive in the long run, firms may need to radically refocus their portfolio resource allocations as technology capabilities and competitor offerings change". Therefore, implementing standard policies of site waste management will greatly assist in a portfolio approach to business, with construction waste having a major detrimental impact on the efficiency of the construction industry (Formoso et al. 2002).

Lingard et al. (1997) argue that effective company site waste management plan s are a key strategy in addressing construction waste, as do Faniran and Caban (1998) and Yates (2013). Osmani et al. (2008) state that the architectural fraternity view of statutory regulation is the critical incentive to site waste reduction, even ahead of creating the appropriate early design. Yuan (2013) notes that ameliorating the regulatory environment for construction waste management is the major strategy for implementing site waste reduction strategies.

Udawatta et al. (2015, pp. 81–82) succinctly summarize the need for a proper waste management plan, supported by corporate policy and strategic regulatory guidelines, when they state that:

> It is necessary to build relationships among stakeholders through high levels of engagement, collaboration and risk sharing, while having adequate supervision with clear instructions to enhance the performance of WM [waste management] practices in construction projects. Indeed, all stakeholders need to be engaged in the WM process, i.e. taking a holistic approach. A proper WM plan can help serve this purpose. Government and company senior management have a responsibility to develop strategic guidelines for WM and to facilitate effective

onsite WM plans by enhancing company policies and regulation relating to construction WM.

In this manner, the following strategies, used from the abridged Yates (2013) front-end site waste minimization strategies model, could be adapted and rolled into the Australian standard code for site waste reduction:

I. Waste generation reduction during project pre-planning to utilize designs that minimize waste using any of the following techniques: precast, prefabrication, pre-assembly and modularization.

II. Considering temporary construction materials versus waste minimization processes. For example, for concrete construction, designers should specify concrete elements of similar dimensions, where practical; use steel shutters on repetitive formwork; ensure formwork is adequately treated and robustly fabricated, to allow re-use; and guarantee that orders are "just in time" to reduce material losses on site.

III. Adopting a structured approach both for engineering design and in the determination of construction methodologies that involves waste minimization strategies.

IV. Evaluating that the mandatory contractor/builder/engineer, consultant/vendor and supplier waste minimization plan has been satisfactorily developed as part of the project execution plan.

V. The contractor/builder/engineer, consultant/vendor and supplier must ensure a minimum quantity of permanent and temporary materials are practically expended in the effective provision of client conforming construction/building products.

A waste minimization strategy in the abridged Yates (2013, p. 293) survey asks: "what techniques are being used that improve resources efficiency, equipment efficiency, material resource efficiency and allow for training of manual labour?" Only 41.7 per cent of respondents for this research survey answered in the affirmative, while Yates' (2013) survey for this question yielded a 56 per cent "yes" response. For example, there is sufficient recent academic literature available to state that Australian worker construction productivity is "sluggish" in improvement over the past 20 years and an estimated 37 per cent behind the USA (Fulford and Standing 2014; Hughes 2014; Langston 2012). As noted earlier in this Sect. 8.6, site waste is considered a major driver of low efficiency in the construction business (Formoso et al. 2002). Therefore, an improvement in site waste could also improve the efficiency of the construction industry; this is a key driver of this research.

Earlier, in this Sect. 8.6, a recommendation was made that the client has an overarching duty of care to ensure regulatory compliance with proposed site waste reduction practices. This would be measured as a triple bottom line effectiveness of the project if the client delivers on these waste minimization initiatives with ensuing positive commercial profit, environmental and social benefits.

The research survey questionnaire Part 3, Question 27, asked respondents whether their respective firms had strategies in place for "innovative designs, construction components, or construction processes, which can be integrated into construction projects to reduce site generated waste". Obviously, there is no process, standard operating procedure or code that can guarantee that innovative practices are adopted on construction projects that reduce site waste.

It is generally recognized that there have been three waves of industrial technological "revolution in the world. The first wave was the Industrial revolution, commenced in the late eighteenth century, with steam power, followed by the second wave, around a 100 years later, during the development of such inventions as the telephone, electric light, aviation and the combustion engine (O'Brien and Marakas 2011). The third wave comprised of the radical and disruptive advancement of information communications technology (ICT) innovation, which commenced with the start of the twenty-first century and arguably ended with the Global Financial Crisis (GFC) (Brynjolfsson and McAfee 2011). The end of this third technological age was due to a contraction of the venture capital available for start-up entrepreneurial high-tech firms; post GFC most ICT technology has re-focussed on step changes, mostly to internet and cloud technologies, rather than radical change (O'Brien and Marakas 2011). Brynjolfsson and McAfee (2011) argue that between 1988 and 2003 the effectiveness of computers increased 43-fold; this was mostly due to advances in algorithms, rather than hardware advances.

Immediately prior to the GFC there was considerable academic literature available surrounding a drive to enhance innovation in the Australian engineering construction industry using information communications technology (Love and Irani 2004; Stewart et al. 2004; Vachara and Walker 2006). These ICT based innovative advances are not as evident in Australian academic literature post GFC. However, a review of respondent comments via thematic analysis has shown that there is still a considerable amount of practitioner interest in efficiency and effectiveness improvements via ICT innovation. There has been a real drive by several of the survey respondents in undertaking the questionnaire on front-end site waste mitigation strategies for each of their particular sectors (planning and scheduling; cost estimating; field project management; and engineering design) to stay abreast of the latest available ICT software, as tools, for each of their respective fields, to effectively and efficiently assist in project management activities and, as a flow on benefit, assist with front-end waste reduction.

Respondents also rated the use of innovation in concrete formwork/falsework design as critical, as well as innovative technologies in general. This made the "innovation" sub-theme the second most critical sub-theme as a driver of front-end waste reduction. "Innovation" is rated as a higher priority as a site waste reduction strategy by Australian respondents, with 48 per cent stressing that "innovation" is considered in their execution processes, while 39 per cent of USA participants in the Yates (2013) research felt that their company considered innovative practices in project implementation.

The author posits that the solution for the provision of the necessary innovation research to improve Australian engineering construction efficiencies via site waste reduction, lies with this industry reaching consensus on where best the limited research funding can be spent. The federal and state governments, construction peak bodies, engineering construction industry and Australian universities need to come together and pool available research resources to develop an innovation strategy and commence work on these innovation priorities. These priorities could be technical, occupational health or industrial relations oriented, for example.

One area that this research has uncovered as being of high importance to the high-rise commercial building respondents, is the urgent need to look at ways and means of reducing packaging and pallet waste on congested urban building sites. This would be an excellent project that a working group from say, the Australian Construction Industry Forum, could work on with members and suppliers, aided by university research, to address this pressing issue. An example of a thoughtful response on packaging and pallet waste reduction is found in Question 27, point two, in which the survey participant notes that:

> Promoting single package bundling of multiple components rather than multiple, individual wrapped components. Delivery and interim storage in containers to provide protection until use—results in reduction of weatherproof packaging.

See Sect. 8.7, "Packaging Material Site Waste Survey Discussion", for further details.

For Question 32, on site waste minimization strategies (which advises that "the contractors/builders, consultants and vendors, need to constructively work with the clients to minimize change orders that make pre-ordered products redundant and suitable only for waste") only 41.7 per cent of respondents answered "yes". In some ways, the development of the means to implement such a change order strategy would prove even more difficult than the successful implementation of the previously reviewed innovation strategy. As noted in Table 7.46, "Discussion of Respondent Answers—Part 3 Site Waste Minimization Strategies Question 26 to Question 34" (see Sect. 7.3.3.1, "Summary of Findings for Analysis of Respondent Responses on Front-End Waste Minimization Strategies), the thematic analysis shows that there is a trichotomy of potentially conflicting interests between the client, contractor and design consultant.

This trichotomy is brought about due to conflicting commercial issues, with each party endeavouring to maximize and/or protect their firm's net profit in the event of a material change to the project's scope. There might be a way to write off this strategy as a commercial matter and not truly the subject of a front-end site waste reduction exercise. However, Sect. 7.2, "Major Sources of Site Waste Findings", of this textbook clearly shows that this research considers client initiated design changes (i.e. change orders) as the greatest potential source of front-end site waste. This argument is supported by the academic literature that suggests design changes are a major source of site waste (Faniran and Caban 1998; Nagapan et al. 2012; Sugiharto 2002).

Fifteen of the 30 respondent comments for Question 32 have been traced to the "communication" sub-theme. Communication refers to the process used to transfer information and influence from one entity to another (Johlke and Duhan 2000, p. 154). More specifically, Krone et al. (1987) suggest that organizational communication can mechanistically be described as having four unique characteristics: frequency of contact; mode and channel used to transfer the message, the message content and or influence strategy; and finally, the direction of communication flow. Schmidts et al. (2001) posit that organizational communication refers to the process whereby individuals and groups transact in a variety of ways, and within different areas, with the aim of implementing organizational goals. However, Brunetto and Farr-Wharton (2004, p. 502) argue that "communication processes affect the identity and organizational climate within an organization, and in turn, the performance of the organization".

Sertyesilisik and Ross (2010) state that the most common causes of client construction project variations are:

I. Scope changes;
II. Client's financial problems;
III. Inadequate client project objectives;
IV. Replacement by client of materials or procedures;
V. Slow client decision-making;
VI. Client drawing or specification changes; and
VII. Obstinate clients.

Hoezen et al. (2006, n.p.) believe that "the efficiency and effectiveness of the construction process strongly depend[s] on the quality of communication" and that given that "construction is a fragmented and dynamic sector" the challenges of communicating effectively are greater than in most other production environments. Contractually driven relationships, conflict, and a lack of mutual respect and trust, all combine to hinder open communication and render the role of the project manager extremely demanding and problematic (Mehra 2018, p. 1).

The importance of a robust project communications plan for the entire project, from inception through to design, execution and operational start up, is vital to successful project implementation and must cover all key internal and external stakeholders (Larson and Gray 2011). Special attention to conflict resolution, both in the communications plan and, of particular importance, in the contractual documentation, must be clearly detailed (Larson and Gray 2011). The client has an obligation to develop precise project objectives and, working from the scoping study initial project phase, have a clear understanding of the project scope; this will minimize design changes during construction (Larson and Gray 2011). Both the engineering consultant and the contractor have an obligation to alert the client to any possible cost, quality and/or schedule implications of possible client directed variations (refer to Sect. 8.5, "Sources of Site Waste Discussion", for the full details, including references).

Table 7.43, "'Short-listed' Critical Themes for Front-End Waste Minimization Strategies" nominated the three key waste-reduction themes as being: (i) "management" (92 respondent comments); (ii) "design" (49); and (iii) "procurement" (19). Table 7.8, "Site waste themes developed from participant survey data on waste sources", nominated (i) "management" (19 respondent comments); (ii) "procurement" (15); and (iii) "design" (nine), as being identified as the three key themes for potential sources of site waste. Undoubtedly, the "management" theme is the leitmotif running through both the sources of waste and the front-end waste minimization strategies research, with a recognition by the respondents that competent management is the critical element for successful waste reduction on Australian projects. The second most important theme for front-end waste minimization strategies thematic analysis was the "design" theme, which was the third most critical theme for sources of site waste.

Because of the high ranking of the most critical source of waste being "design changes" in this research and other academic papers (Faniran and Caban 1998; Nagapan et al. 2012; Sugiharto et al. 2002), the author has ranked "design" as the second most critical theme. This is especially so because the highest ranked sub-theme for waste minimization strategies was the "front-end engineering design" sub-theme. "Procurement" is nominated as the third most critical theme in this research, coming in at that ranking for site waste minimization strategies. An "integrated logistics chain" was the most commented upon sub-theme and the "procurement" theme was third for sources of site waste, for which the predominant sub-theme was "quality process shortfalls".

A review of Chap. 7, "Findings", of sub-themes, indicates the following:

I.   The "front-end engineering design" (FEED) sub-theme was the most important sub-theme for site waste minimization strategies research and was rated second for the sources of waste sub-theme. This sub-theme is part of the over-arching umbrella of the "design" theme.

II. "Innovation" was deemed the second most important sub-theme for front-end site waste minimization. This sub-theme can be tied back to the "management" theme as the critical driver in allowing sufficient capital for implementing these innovations.

III. The "client responsibility" sub-theme had 33 respondent comments and was the third most critical theme for site waste minimization strategies. This theme was equal third for sources of site waste, with three respondent comments expressing the view that "client responsibility" is the main respondent concern surrounding the "management" theme.

IV. The "quality process shortfall" sub-theme was the most critical sub-theme for potential sources of site waste. However, it was rated extremely low for the "front-end site waste minimization" sub-theme. This sub-theme was a catch-all for non-compliances in project execution plan processes that ranged from poor material take-offs to sub-standard site warehouse facilities for project material storage.

Academic literature suggests that a design focused method of construction waste reduction offers the most cost-effective and flexible minimization strategy, which is in keeping with the findings of this research (Akinade et al. 2018; Ajahi et al. 2016; Faniran and Caban 1998). Current site waste minimization design tools have limited capabilities and these tools are incompatible with Building Information Modeling (BIM) software (Akinade et al. 2018; Cheng and Ma 2013).

BIM is a process involving the generation and management of digital representations of physical and functional characteristics of a facility and this software is used to plan, design, construct, operate and maintain diverse physical infrastructure (Eastman et al. 2011, Preface, p. i).

Akinade et al. (2018) argue that further research needs to be undertaken to determine stakeholder expectations for the use of BIM as a design tool in minimising construction waste. This approach ties in the two critical sub-themes, above, for "FEED" and "innovation".

Throughout this textbook, the author has illustrated in several instances that any respondent remarks on the recycling and re-use of materials has been denoted as "not in the research scope" for this book. Survey participants were advised in the questionnaire and informed individually by email that front-end site waste minimization strategies would be the only subject of the research. However, as a conclusion to the subject, it is reiterated that the provision of adequate recycling resources is in fact a front-end site waste reduction practice. This is because adequate laydown on-site for temporary waste storage, ample salvage bins, and sufficient personnel to sort materials is necessary to ensure that suitable materials go to recycling/re-use. Otherwise, as succinctly mentioned by one respondent, all other site waste is captured for transportation to bulk landfill disposal.

As a conclusion to this Sect. 8.6 discussion, it is useful to re-visit the details of a University of Exeter report on construction waste minimization (Vowles 2011), outlined previously in Sect. 4.3, "Broader Community Benefit", of this textbook. This University of Exeter guideline paper suggests that by adopting a set of 10 strategies, project capital costs can be reduced by two per cent and they have integrated this process in their capital works program (Vowles 2011). These 10 strategies are quoted as follows:

The proposed actions are therefore listed in chronological order as the project progresses:

(a) Provide regular training for EDS staff to update on issues with construction waste and means to minimise it and embed this within the EDS ISO9001 QA system.

(b) Include within the project brief and the tender documents for the consultant's services and construction works, the University's aim and requirements to minimise the environmental impact caused by any construction works.

(c) The project team will consider means to minimise construction waste from the inception to the completion of the project, through the initial brief, design process, materials selection, construction techniques and operational methods.

Examples of specific requirements include:

(d) Inclusion within the Scope of Services for the appointment of designers, a requirement to comply with the 'Designing out waste: a design team guide for buildings' guideline. This guide encourages designers to design out waste by signing up to the five key principles design teams can use during the design process to reduce waste, namely to:
- Design for re-use and recovery.
- Design for off-site construction.
- Design for materials optimization.

- Design for waste efficient procurement.
- Design for deconstruction and flexibility.
- Production of whole life cycle costs for key structural and services elements of construction projects during the design process.
- Inclusion within the EDS ISO9001 QA system, as part of the briefing process for refurbishment/alterations projects, a requirement for the Project Manager/Designer to question the need to replace fittings and fixtures in buildings, where it is economically viable to retain them (Vowles 2011, pp 1–3).

These above 10 points from the University of Exeter paper (Vowles 2011) have essentially validated several of the major issues uncovered in this research.

Briefly, this research is in alignment with the following common findings:

   I.   The importance of an overarching regulatory code of practice.
  II.   The introduction of a cultural change to the construction industry to focus on site waste management and use training to continually reinforce this cultural change.
 III.   Constant constructive communication between the client, contractor and design consultant through the whole of the project life cycle will assist in reducing site waste that occurs from scope changes.
 IV.   The importance of thoughtful front-end engineering design, with a focus on waste reduction.
  V.   Client, contractor and design consultant should all have project waste management plans that reside under an overarching regulatory code.
 VI.   Efficient procurement processes are important.
VII.   Offsite fabrication will maximize waste reduction.

8.7 Packaging Material Site Waste Survey Discussion

In Sect. 7.3.5, "Consolidated Summary of Front-End Site Waste Minimization Strategies Findings", it was recommended that further discussions should be undertaken to determine the viability of preparing a supplementary survey question on front-end strategies to reduce packaging and pallet waste. After receiving written advice from the SCU ethics committee chair that a simple survey for a select cadre of participants was a viable research methodology, the author prepared a survey with input from the co-authors. The purpose of this exercise was to determine to what degree the Australian engineering construction industry was addressing this issue of packaging and pallet waste.

The survey was distributed to a discrete sample batch of engineering construction industry personnel from the 102 participants who were invited to complete the three part questionnaire, which was the main qualitative methods data accumulation tool adopted in this research. A sample pool of 50 participants was selected from the original pool of 102 practitioners, with 17 respondents filling in the single survey question and returning their answers via the internet.

Participants were asked to answer "yes", "no", or "don't know", and if the respondents answered in the affirmative, they were asked to elaborate further on what strategies were being adopted to minimize this waste. Of the 17 participants who replied to this survey, only two respondents, both clients, indicated that provisions were made with their suppliers to reduce packaging and pallet waste as a condition of the shipment of physical materials to the site. The balance of responses were 12 "no allowance made for packaging waste reduction" and three "don't know" responses. Therefore, the author is confident that saturation has been reached. Creswell (2013) states that, typically, somewhere between eight to 30 surveys are required to attain saturation. In any event, with a response rate of "no" for 12 participants and only "yes" from two participants, from a sample pool of 17 engineering construction industry participants, there is little doubt that continued polling of this question with a larger participant pool would yield similar results (Glaser and Strauss 1967).

Pursuant to the above, this study suggests that there is little being done in the construction industry to reduce packaging and pallet waste on construction projects.

8.8 Variances in Literature on Site Waste Definition Discussion

The peak Australian construction body, the Australian Construction Industry Forum (ACIF), argues that over 30 per cent of the Australian engineering construction industry's efforts are misdirected and that reducing construction wastage by a third would save this industry in excess of A\$10 billion per annum (Barda 2013). Josephson and Saukoriipi (2007) argue that the cost of waste totals between 30 and 35 per cent of a project's contract sum on building projects in Sweden. Physical construction waste amounts to around 13 per cent, on average, of a project's contract sum on building projects in Scotland (Zero Waste Scotland 2017). Depending on the type of project, this textbook notes that physical construction waste in Australia adds, on average, an extra 6.5 per cent to the project's cost (Doab 2018). The Master Builders Association of Victoria states that physical site waste costs add an extra 10 per cent onto their projects (Doab 2018; MBAV 2004). The wide divergence between these research findings and the much higher construction waste numbers cited by the ACIF (Barda 2013) and Josephson and Saukoriipi (2007) can be readily explained.

Construction waste can be classified into physical and non-physical groups (Alarcon 1994). This textbook examines the sources of physical waste and the strategies to reduce front-end physical waste. Physical construction waste is defined as waste which arises from construction, renovation and demolition activities including land excavation or formation, and civil or building construction, with the waste product comprising of physical inert temporary and/or material construction materiel, including but not limited to bricks, concrete debris, timber off-cuts, tiles, glass, vegetation waste from site clearing, steel off-cuts, bricks and blocks, plasterboard, plastic wrapping and pallets (Alarcon 1997; Ferguson 1995; Formoso et al. 1999; Yates 2013).

Nagapan et al. (2012) defines non-physical waste as that which normally occurs during the construction process. By contrast with material waste, non-physical wastes are related to time and cost overruns for construction projects due to labour and equipment resource productivity inefficiencies (Nagapan et al. 2012). Physical construction waste is the subject of this detailed research textbook.

However, both the ACIF definition by Barda (2013) of the amount of waste on a project and the Josephson and Saukoriipi (2007) definition of the cost of waste to a project go far beyond the scope definition of "physical construction waste" studies in this research.

The ACIF posits that 30 per cent of project waste is typically caused by such matters as:

I. Several iterations of drawings (between three to 10) as a result of multi-stakeholder inputs and re-design to address client budget cost reduction matters in the preliminary project phases;
II. Up to 40 per cent of drawings by design consultants are not required for use by trade contactors for construction;
III. Poor drawing coordination between design consultants and trade contractors; and
IV. Th cost of defective work (Barda 2013).

The ACIF noted that a 2004 Lean Construction Institute study concluded that 57 per cent of time, effort and material investment in the construction industry adds no value to the final product, compared to only 26 per cent in the manufacturing industry (Barda 2013).

Josephson and Saukoriipi (2007) state that several of the Swedish contractor participants noted that, in the past 10 years, researchers have advised that their waste costs have progressively risen from five per cent to 10 per cent and are currently 35 per cent of the contract value. However, Josephson and Saukoriipi (2007) also note that the issue is one of changing definitions, perspectives, viewpoints, scope and research methods surrounding what constitutes project waste. The Josephson and Saukoriipi (2007) definition of project waste, which costs between 30 to 35 per cent of a project's capital cost, is noted as follows:

I. Material waste from visible and hidden defects such as the costs of checks, inspections, insurances, theft and vandalism;
II. Inefficient utilization of plant, equipment, material and labour;
III. Work related injuries, including rehabilitation and illness; and
IV. Expensive and time consuming regulatory processes, such as planning/building approval, environmental approval, and land acquisition and resumption (Josephson and Saukoriipi 2007).

Pursuant to the above, it can be seen that the three different definitions of project waste vary markedly between this textbook, the ACIF definition (Barda 2013) and Josephson and Saukoriipi's (2007) viewpoint. This may explain the large differences in waste costs and waste quantities.

References

Ajayi, S. O., Oyedele, L. O., Kadiri, K. O., Akinade, O. O., Bilal, M., Owolabi, H. A., et al. (2016). Competency-based measures for designing out construction waste: Task and contextual attributes. *Engineering, Construction & Architectural Management Journal, 23*(4), 464–490.

Akinade, O. O, Oyedele, L. O., Ajayi, S. O., Bilal, M., Alaka, H. A., Owolabi, H. A., et al. (2018). Designing out construction waste using BIM technology. Stakeholders' expectations for industry deployment. *Journal of Cleaner Production, 180*, 375–385.

Alarcon, L. F. (1997). *Lean construction processes*. Chile: Catholic University Press.

Alarcon, L. F. (1994). Tools for the identification and reduction of waste in construction projects. In L. Alarcon (Ed.), *Lean construction*. Netherlands: A.A., Balkema Publishers.

Amado, M, Ashton, K, Ahston, S, Bostwick, J, Clements, G, Drysdale, J, Francis, J, Harrison, B, Nan, V, Nisse, A, Randall, D, Rino J, Robinson, J, Snyder, A, Wiler, D., & Anonymous (2012). *Project management for instructional designers*. Press Books, USA.

Ameh, O. J., & Daniel, E. L. (2013). Professionals' views on material wastage on construction sites. *Technology Management Construction, 5*, 747–757.

Argote, L., Ingram, P., Levine, J., & Moreland, R. (2000). Knowledge transfer in organizations. *Organizational Behaviour and Human Decision Processes, 82*(1), 1–8.

Australian Constructors Association (ACA). (2017). *Australian constructors association construction outlook at June 2017*, Australian Constructors Association, viewed June to September 2017. http://www.constructors.com.au/wp-content/uploads/2017/08/Construction-Outlook-June-2017.pdf.

BHP Billiton.(2007). *Project manual for capital investment projects*. BHP Billiton Project Management Centre, Australia.

Barda. (2013). *Stop the waste—30% is too much*, viewed March 2018. https://www.acif.com.au/news/opinion/stop-the-waste—30-is-too-much.

Braun, V., & Clarke, V. (2006). Using thematic analysis in psychology. *Qualitative Research in Psychology, 3*(2), 77–101.

Brunetto, Y., & Farr-Wharton, R. (2004). Does the talk affect your decision to walk: A comparative pilot study examining the effect of communication practices on employee commitment past-managerialism. *Management Decision, 42*(3/4), 579–600.

Brynjolfsson, E., & McAfee, A. (2011). *Race against the machine*. USA: W.W. Norton.

Chandler, D. (2016). Construction waste is big business, *Construction news, 12* October, viewed May 2017–May 2018. https://sourceable.net/construction-waste-is-big-business/.

Charmaz, K. (2006). *Constructing grounded theory: A practical guide through qualitative analysis*. Thousand Oaks, USA: Sage.

Cheng, J. C. P., & Ma, L. Y. (2013). A BIM-based system for demolition and renovation waste estimation and planning. *Waste Management Journal, 33*, 1539–1551.

Coakes, S. J., & Ong, C. (2011). *SPSS: Analysis without anguish version 18.0 for windows*. Milton, Queensland: Wiley.

Creswell, J. W. (1998). *Qualitative inquiry and research design: Choosing among five traditions*. Thousand Oaks, USA: Sage.

Creswell, J. W. (Ed.). (2013). *Qualitative inquiry and research design—choosing among five approaches*. USA: Sage Publications.

Department of Industry, Innovation and Science (DIIS). (2017). *Building/construction fact sheet*. DIIS, South Australia.

Doab, M. (2018). Average cost of construction site waste on major australian projects. *Doab Estimation Enterprises Proprietary* D'Est Estimating Model.

Eastman, C., Tiecholz, P., Sacks, R., & Liston, K. (2011). *BIM Handbook: A Guide to Building Information Modeling for Owners, Managers, Designers, Engineers and Contractors*. New Jersey, USA: John Wiley.

Eisenhardt, K. M., & Zbaracki, M. J. (1992). Strategic decision-making. *Strategic Management Journal, 13,* 17–37.

Faniran, O. O., & Caban, G. (1998). Minimizing waste on project construction sites. *Engineering Construction and Architectural Management, 17*(1), 57–72.

Ferguson, J. (1995). *Managing and minimizing construction waste: A practical guide.* London, UK: Institute of Civil Engineers.

Formoso, C. T., Isatto, E. L., & Hirota, E. (1999, July). *Method for waste control in the building industry.* Paper presented to the 7th Annual Conference of the International Group for Lean Construction, University of California, Berkley.

Formoso, C. T., Soibelman, L., Cesare, De, & Isatto, E. L. (2002). Material waste in building industry: Main causes and prevention. *Journal of Construction Engineering Management, 128,* 316–325.

Fulford, R., & Standing, C. (2014). Construction industry productivity and the potential for collaborative practice. *International Journal of Project Management, 32*(14), 215–326.

Gavilan, R. M., & Bernold, L. E. (1994). Source evaluation of solid waste in building construction. *Journal of Engineering and Management, 120*(3), 536–555.

Glaser, B., & Strauss, A. (1967). *The discovery of grounded theory.* Hawthorne, NY: Aldine Publishing Company.

Grossman, H. A. (2005). Refining the role of the corporation: The impact of corporate social responsibility on shareholder primacy theory. *Deakin Law Review, 10*(2), 572–597.

Hickey, J. (2015). *Landfilling construction waste in Australia,* viewed April to December 2017. https://sourceable.net/landfilling-construction-waste-in-australia/.

Hoezen, M., Reynen, I., & Dewulf, G. (2006). *The problem with communication in construction.* Paper presented at the International Conference on Adaptable Building Structures, Eindhoven, Netherlands.

Hughes, R. (2014). A review of enabling factors in construction industry productivity in an Australian environment. *Construction Innovation Journal, 14*(2), 210–228.

Johlke, M., & Duhan, D. (2000). Supervisor communication practices and service employee outcomes. *Journal of Service Research, 3*(2), 154–165.

Josephson, P., & Saukoriipi, L. (2007). *Waste in construction projects: Call for new approach.* Sweden: Chalmers University of Technology Publication.

Kester, L., Hultink, E., & Griffin, A. (2013, June). Empirical exploration of the antecedents and outcomes of NPD portfolio success. In *International Product Development Management Conference*, Paris, France.

Kish, L. (1995). *Survey sampling.* USA: Wiley.

Krone, K., Jablin, F., & Putnam, L. (1987). *Handbook of organizational communication: An interdisciplinary perspective.* California: Sage Publishers.

Langston, C. (2014). Construction efficiency: A tale of two countries. *Energy, Construction & Architectural Management, 21*(3), 320–325.

Langston, C. (2012). *Comparing international construction performance.* Australia: Institute of Sustainable Development & Architecture, Bond University Publishing.

Larson, E., & Gray, C. (Eds.). (2011). *Project management: The management process.* New York: McGraw-Hill/Irwin, The McGraw-Hill Companies Inc.

Lewin, K. (1947). *Field theory in social science.* London: Social Science Paperbacks.

Lingard, H., Graham, P., Smithers, G. (1997). Waste management in the Australian construction industry. In *13th Annual Kings College Cambridge UK ARCOM Conference.* Cambridge University.

Love, P., & Irani, Z. (2004). An exploratory study of information technology evaluation and benefits management practices of SMEs in the construction industry. *Information & Management, 42*(1), 227–242.

Mason, M. (2010). Sample size and saturation in PhD studies using qualitative interviews. *Forum for Qualitative Social Research, 11*(3), 1–19.

Mehra, S. (2018). *Importance of communications in the construction industry*, viewed November 2017. https://www.scribd.com/document/7875707/Project-Communication-Summary-by-Sachin-Mehra.

Moody, R. (2013). *Rocks and hard places—Globalization of mining*. London, UK: Zed Books.

Morse, J. M. (1995). The significance of saturation. *Journal of Qualitative Health, 5*(2), 147–149.

Nagapan, S., Rahman, I., & Asmi, A. (2012). Factors contributing to physical and non-physical waste generation in construction, *International Journal of Advances in Applied Sciences, 1*(1), 1–10.

Osmani, M., Glass, J., & Price, A. (2008). Architects' perspectives on construction waste reduction by design. *Waste Management Journal, 28,* 1147–1158.

O'Brien, J., & Marakas, G. (Eds.). (2011). *Management Information Systems*. New York: McGraw-Hill, Irwin.

Pope, C., Ziebland, S., & Mays, N. (2000). Analysing qualitative data. *British Medical Journal, 320,* 114–116.

Schmidts, A., Pruyn, A., & von Riel, (2001). The impact of employee communication and perceived external prestige on organization identification. *Academy of Management Journal, 44*(5), 1051–1062.

Sertyesilisik, B., & Ross, D. (2010). Variations and change orders for construction projects. *Journal of Legal Affairs and Disputes, 2*(2).

Slaper, T., & Hall, T. (2011). The triple bottom line: What is it and how does it work. *Indiana Business Review, 8*(1), 4–8.

Spivey, D. (1974). Construction solid waste. *Journal of the American Society of Civil Engineers, Construction Division, 100,* 501–506.

Stewart, A., Mohamed, S., & Marosszeky, M. (2004). An empirical investigation into the link between information technology implementation barriers and coping strategies. *Australian construction industry, Construction Innovation, 4*(3), 155–171.

Stiles, W. (1993). Quality control in quantitative research. *Clinical Psychology Review, 13,* 593–618.

Strauss, A., & Corbin, J. (1998). *Basics of qualitative research: Techniques and procedures for developing grounded theory*. Thousand Oaks, USA: Sage.

Sugiharto, A., Hampson, K., Sherid, M. (2002). *Non value adding activities in Australian construction projects*. Paper presented to the International Conference for Advancement in Design, Construction, Construction Management and Maintenance of Building Products, Griffith University, Australia.

Sundqvist, E., Backlund, F., & Chroneer, D. (2014). What is project efficiency and effectiveness? *Procedia—Social and Behavioral Sciences, 119,* 278–287.

Swinburne University. (2016). *How to write a research question*, SU School of Electrical Engineering, viewed March to October 2017. https://www.youtube.com/watch?v=lJS03FZj4K.

Tonks, G. (1981). *The establishment, operation and subsequent closure of the Bougainville copper mine: A case study in international management*. (PhD Dissertation) University of Tasmania, Tasmania, pp. 1–475.

Udawatta, A., Zuo, J., Chiveralls, K., & Zillante, G. (2015). Improving waste management in construction projects: An Australian study. *Resources, Conservation and Recycling, 101,* 73–83.

Vachara, P., & Walker, D. (2006). Information communication technology (ICT) implementation constraints: A construction industry perspective. *Engineering, Construction and Architectural Management, 13*(4), 364–379.

Vowles, F. (2011). *Construction strategy*, University of Exeter, viewed November to December 2017. https://www.exeter.ac.uk/media/universityofexeter/campusservices/sustainability/pdf/construction-strategy.pdf.

Wacaster, S. (2010). *Mineral commodity summaries*, US Geological Survey, viewed March 2018. https://minerals.usgs.gov/minerals/pubs/mcs/2010/mcsapp2010.pdf.

Yates, J. (2013). Sustainable methods for waste minimisation in construction. *Construction Innovation, 13*(3), 281–301.

Yuan, H. (2013). Critical management measures contributing to construction waste management: Evidence from construction projects in China. *Project Management Journal, 44*(4), 101–112.

Zero Waste Scotland. (2017). *The cost of construction waste*, Government of Scotland, viewed February to June 2017. http://www.resourceefficientscotland.com/content/cost-construction-waste.

Chapter 9
Conclusions

9.1 Preamble to Key Conclusions

It is appropriate to preface this subsection, summarizing the key conclusions of this textbook, with a recapitulation of exactly why the independent selection by two review panels of nine eminent engineering constructions peers from the engineering construction business and academic world, ultimately ended in the selection of the most appropriate topic for investigation. The original purpose of this study was to identify eight possible methodologies that would have the potential to improve the efficiency and effectiveness of the Australian engineering construction industry, with a view to selecting the most appropriate methodology for detailed research. After a rigorous two-stage review supervised by an independent facilitator, a panel of eminent peers adopted a decision analysis process to select a four option "short-list" for further review. This enabled a second peer review panel to select a single "Go Forward" option for further detailed research.

The peer review panel directed that a number of questions be answered before the selection of the "Go Forward" case. After further research of available academic, trade and professional literature, suitable responses were provided by the author and the panel recommended that detailed research be undertaken on "*Front-end Waste Minimization Strategies on Australian Construction projects*". This selection by the independent peer review panel of this option has proven most prudent.

Australian residential and low-rise commercial building project costs generally make a 10 per cent allowance to account for physical construction material waste. Major Australian heavy construction projects generally estimate for a 6.5 per cent cost allowance for material wastage; these figures are consistent with overseas academic literature (Doab 2018; Jain 2012; Josephson and Saukoriipi 2007; MBAV 2004). Depending on the type, size and location of construction projects in Australia, anything between eight to 30 per cent of construction materials are wasted and disposed of as landfill or recycled in Australia, with 13 per cent being an average estimate of the material wastage on construction projects (Core Logic 2016; Zero Waste Scotland 2017).

© Springer Nature Switzerland AG 2019
P. G. Rundle et al., *Effective Front-End Strategies to Reduce Waste on Construction Projects*, https://doi.org/10.1007/978-3-030-12399-4_9

A University of Exeter report on construction waste minimization suggests that by adopting a set of 10 strategies, project capital costs can be reduced by two per cent (Vowles 2011). These 10 University of Exeter waste minimization strategies are in basic alignment with the findings of this research synthesis of the extensive data from a survey of over 50 construction industry participants, using two adapted seminal models to identify sources of site waste and validate the applicability of front-end site waste reduction strategies (Faniran and Caban 1998; Yates 2013). It has been estimated that construction waste amounted to A\$2.8 billion, annually, for all residential building construction in Australia alone, excluding high-rise apartment construction (Chandler 2016).

Carrying this simple logic forward, there could be a potential savings of two per cent. Using data from the peak industry body, the Australian Construction Industry Forum, turnover for all construction in Australia, in 2016, was A\$218 billion, which, by sector, includes residential building construction (\$96 billion), non-residential building construction (\$37 billion) and engineering construction (\$85 billion) (ACIF 2017). A saving to the nation's engineering construction industry of two per cent, or, A\$4.36 billion is possible by following these front-end waste minimization strategies that require no innovative technology or large capital infrastructure expenditure such as recycling and recovery facilities. Even large savings in front-end waste reduction would be enjoyed by the Australian heavy engineering construction and commercial building sectors. Public and private sector capital works clients; Australian home-owners; public housing; negative gearing for investors not paying built-in site waste costs; and contractor and builder stakeholders would all benefit from these savings. Engineering consultants and building product suppliers would also benefit, as the residual available investment capital would likely create employment and increase GDP by encouraging public, private and citizen expenditure.

Waste Management World's 2016 "*State of the Nation*" reported that:

constructionand demolition (C&D) waste (typically timber, concrete, plastics, wood, metals, cardboard, asphalt and mixed site debris such as soil and rocks) comprises approximately 40 per cent of Australia's total waste generation (Waste Management World 2016, n.p.).

A 2011 Hyder report commissioned by the Australian Federal Government shows that construction and demolition waste accounted for 31 per cent of landfill waste in Australia (Hyder Consulting 2011). Chandler (2016) states that the Australian construction industry is the largest "offender" of any industry, accounting for a third of all waste sent to landfill.

Australian Federal Government data on landfill sites notes that there is a minimum of approximately 600 medium to large facilities currently in operation, and while the total number of landfill facilities are unknown, there could be as many as 2000 unregistered and unregulated landfill sites in Australia (Waste Management World 2016). The Australian construction industry is a major user of private landfill sites, and it is concluded that measures to reduce front-end construction waste could reduce both registered and unregistered landfill footprints, which would provide considerable social and environmental benefits.

Chandler (2016) summarizes that the construction industry is the globe's major consumer of raw materials internationally, but that around only one-third of construction and demolition waste is recycled or reused, as per a recent World Economic Forum report. Australia is one of the 10 highest producers of solid waste in the OECD (Chandler 2016). Similar to Australia, solid waste in the USA typically comprises around 40 per cent of construction and demolition waste per annum (Chandler 2016). Any reduction in construction material waste shall improve the productive use of Australia's finite volume of raw materials.

Australian governments have declared recycling targets to divert waste away from landfill disposal, so as to recover these materials for recycling and/or re-use, with a caveat that effectively all recycling comes at an intrinsic economic cost to waste generators (be they industrial or domestic) (Ritchie 2016). Though Ritchie (2016, n.p.) argues that:

> [g]enerally, metals, paper and cardboard and plastic (in sufficient quantities), are commercially viable recyclables. Almost all other recycling in Australia is subsidised by someone via gate fees, grants or the like. That includes most household, construction and commercial waste. Unfortunately, there is no free lunch in recycling.

Australia sends approximately 30 per cent of its recyclable waste offshore to China; however, China has banned the import of foreign waste since the beginning of 2018, throwing the Australian recycling industry into turmoil (Lasker 2018). This is coupled with a parallel reduction by 30 per cent in bulk recycled waste products like plastics, paper, cardboard and glass (Lasker 2018). This may have a negative cost impact on the Australian construction industry, which countrywide, recycles on average 58 per cent of construction waste (Hickey 2015).

New South Wales charges a punitively high levy of A$133.10 per tonne of waste for landfill waste to encourage recycling, while Victoria imposes a landfill levy of A$60.52 per tonne. South-east Queensland has an average levy of only A$30 per tonne of waste to landfill, which has resulted in the significant disposal of construction waste and other large volumes of waste from New South Wales and Victoria across the border to Queensland, with an estimated total of 875,000 tonnes of waste disposed in Queensland landfill sites by these other two states in 2014 and 2015, which created an extra 15,000 truck movements along the Pacific Highway (Ritchie 2016). These cross-border truck transportation waste disposal activities are causing extra pavement wear and increasing accident risk potential. Front-end waste minimization, as recommended in this book, will result in a significant decrease in landfill footprint, without the need to apply punitively high disposal costs to encourage back-end recycling.

There shall always be a significant place in the handling of construction site waste as a back-end waste treatment option, when commercially viable, and there will always be a need for government regulation by imposition of a just levy, which will be a cost passed on to the capital works project client by the contractor/builder (Chandler 2016; Hickey 2015; Ritchie 2016). However, because the Western Australian government is the only Australian government that returns a portion of landfill levies (25 per cent) back into waste management, it has been concluded by the author that there is a second option, other than returning landfill levies back into consolidated revenue (Hickey 2015). All Australian states and territories could spend these landfill levies on a program of improvements to recycling infrastructure, supported by a review of the recyclable construction material market development and levy-funded training schemes aimed at the construction industry and not just politically motivated household waste focused education, as is now the case (Hickey 2015). This is even more pertinent, in view of the Chinese ban on the export of Australian recyclable waste, the collapse of recyclable commodity prices, coupled with the need for a general upgrade of below par Australian recycling facilities (Lasker 2018).

A series of recent 2017 and 2018 Australian Fairfax newspaper articles and Australian Broadcasting Corporation (ABC) television documentaries contend that a significant volume of domestic and industrial waste, diligently sorted by waste-producing stakeholders, is disposed of by the statutory authorities into landfill (cited in Lasker 2018). Pursuant to the above, the front-end construction waste minimization processes, nominated in this textbook, become even more prescient as the recycling business enters a period of flux. The front-end approach to construction waste reduction bypasses the carbon emissions caused by recycling transportation and processing, along with the alleviation of general traffic congestion by waste and bin-handling operations.

The total benefits attributable to a reduction in construction material site wastage shall include, but not necessarily be limited to:

I. Carbon footprint reductions via manufacture transportation and disposal/recycling of significantly less construction materials;
II. Savings in energy and water via significantly less construction material manufacturing and recycling;
III. Ecological benefits for the groundwater and aquifer systems, due to landfill reductions;
IV. Savings in the cost of transportation, procurement and storage and landfill;
V. Environmental benefits to the broader community via pollution reduction and less impact on the environmental footprint via a reduction of landfill sizes, and a reduction of unregulated and possibly contaminated landfill sites; and
VI. Savings in rare and valuable irreplaceable raw resources, such as lime, timber, sand, clay, rock, gypsum, coking coal for steel manufacture, titanium, copper, bauxite and iron ore.

Furthermore, a reduction in residential building industry costs of A$58 million due to site waste reduction may assist home buyers via a reduction in construction costs. Contractor, builder and client stakeholders shall all benefit, commercially, from the project capital cost reduction due to front-end site waste minimization initiatives. The benefits to the community at large are clear; these range from a decrease in unregistered potentially contaminated landfill sites, to lower cost public housing and an increase in employment through an expansion in construction activities, as waste reduction savings are ploughed back into capital works programs.

Refer to Appendix C, which details the conclusions for the research methodology and the void in the literature.

9.2 Sources of Site Waste Conclusions

This research has determined that the six major potential sources of site waste are:

   I. Client-initiated design changes;
  II. Design and detail errors;
 III. Packaging and pallet waste;
 IV. Procurement ordering and take-off errors;
  V. Lack of on-site materials planning and control; and
 VI. Improper material storage.

The conclusion that can be made from this research is that the Faniran and Caban (1998) sources of site waste identification model, duly adapted with the refinements suggested by the author in Chap. 10, "Recommendations", is a suitable tool for similar future research. The primary benefits of using this model are that research comparisons can be readily made between synthesized data sets from various studies. Another benefit of using this model was that there were no other sources of waste identified between the Faniran and Caban (1998) study and this 2017 study and that the findings on the most likely sources and least likely sources of construction site waste are similar in nature.

In a review of two other Australian research studies on construction waste, one approach looked at human factors impacting sources of waste, while the other paper largely focused on the types of design issues impacting upon the causation of site waste (Sugiharto et al. 2002; Udawatta et al. 2015). Both studies provided valuable data that could be correlated with this research. However, the author is satisfied that for their research purposes, the adapted Faniran and Caban (1998) model was the correct choice, both for data validity purposes and because the model output showed that none of the 53 respondent samples provided other possible sources of waste when questioned in the survey. In summary, based upon a comparison between this research and the seminal study undertaken by Faniran and Caban (1998) the main sources of construction site waste in Australia have remained static for 20 years.

The top three sources of site waste in this research that indicates a correlation with the Faniran and Caban (1998) research are:

   I. Design changes (initiated by client);
  II. Packaging and pallet waste (and other non-consumables); and
 III. Design and detailing errors.

Note that this research divided "design changes" between "client initiated" and "contractor initiated" sources. However, the Faniran and Caban (1998) model adopted only a "design changes" source of waste.

When comparing the data from this research and the Faniran and Caban (1998) Australian data set with both surveys using the same questionnaire template, an analysis of the four least likely sources of construction waste for each data set indicates that there are two common results, which are:

  I. Site accidents; and

  II. Criminal waste (from pilfering and vandalism).

Through a triangulation check on this research, for data validity, it was found that the results from an Australian paper by Sugiharto et al. (2002) also determined that design changes and design/detailing errors were two major potential sources of site waste; this is in line with this research and the Faniran and Caban (1998) research. Similarly, Nagapan et al. (2012) listed design changes as a major cause of site waste and their global literature review indicated that there was a low propensity for criminal activity and site accidents to be significant contributors to potential site waste.

In summary, it can be concluded that the Australian engineering construction industry is in a strong position, armed with reliable information from a 20-year period on the sources of construction waste and now needs to attack these sources as a means of reducing waste.

A thematic analysis was carried out on Question 25 of this survey, which asked survey participants to list any potential sources of waste other than those noted in the Faniran and Caban (1998) sources of site waste identification model. An abridged form of thematic analysis was undertaken of the results from the Faniran and Caban (1998) sources of site waste model findings from this research. Nagapan et al. (2012) schedule of themes for sources of site waste, prepared via a worldwide literature review, was applied to these model results and the appropriate sub-themes were synthesized by the author from the typology of each potential source.

A thematic analysis of respondent comments found that the 53 participants in this survey indicated that the major themes, in descending order, for sources of site waste were:

   I. Management;

  II. Procurement;

 III. Design; and

 IV. Workers.

A thematic analysis to synthesize these respondent comments found that the major sub-themes, in descending order were:

   I. Quality process shortfalls;

  II. Front-end engineering design (FEED);

 III. Client responsibility; and

 IV. Work culture.

The conclusion drawn from the thematic analysis was that the "management" theme (more particularly, the "client management" sub-theme) was a primary driver of potential sources of site waste. As the generator and funder of a project, the client retains the ultimate responsibility for project approval and that includes the processes for how site waste is handled on their project, through development and implementation of the contract terms and conditions with the designer, supplier and the contractor. If one views three of the top sources of waste from the research survey using the Faniran and Caban (1998) model, namely (i) procurement ordering and take-off errors; (ii) lack of on-site materials planning and control; and (iii) improper material storage, it is obvious that the "procurement" theme is a powerful influencer of several major sources of site waste.

A sub theme that breaks these down, makes of procurement into a sub-set and points the way to a solution is the "quality process shortfall" sub-theme. Procurement take-off errors, lack of a site procurement plan and a failure to address site material matters, such as secure and weatherproof warehousing and materials handling, are all issues that can be handled, and material site wastes may be reduced by developing and following appropriate procurement standard operating procedures. The cited academic literature in this textbook has argued that the necessary costs to implement proper procurement processes, borne by the client through the construction contract, will be offset by the significant savings in site material waste reduction, which can run from between 21 and 30 per cent of project cost over runs.

Further literature review, upon the completion of Chap. 7, "findings", has led to the author's conclusion that "client-initiated design changes" can be minimized by the correct life cycle planning of a project, commencing at inception when a client nominates their business objectives for a project and outlines the project scope of the works aligning with the client's project objectives. This life cycle should run as follows: (i) the preliminary phase of a project comprising a scoping study to develop project scoping requirements; (ii) a concept study to determine a broad basis for whether a potential project is viable; (iii) a pre-feasibility study to select the most feasible option; and (iv) a feasibility study that provides sufficient engineering designs, project cost estimates and a project schedule, along with other deliverables such as environmental, regulatory and financial modelling, to determine the robustness of the proposal.

It is possible to cross-link the "client initiated design changes" number one source of site waste, derived in this research, with the "design" theme, but also with the "front-end engineering design" (FEED) sub-theme. It is critical that design consultants involve clients from inception with design development and that the client can validate that their business objectives are being met by the project and that the design complies with the client's understanding of the project's scope. In this manner, material waste, due to client-directed design revisions because of requested scope changes, can be minimized.

It is concluded that the reflexivity journal, used as a tool throughout this thematic analysis, proved valuable as a means of mitigating confirmation bias from the author and for maintaining an objective view during the data evaluation process. It has been a valuable lesson for the author that the 53 respondents, including a cadre of 16 engineering construction company directors, had the objectivity to acknowledge that the prime responsibility for construction waste control resides with management and not the worker. This was validated in this research by the "management" theme being the top theme across both the sources of waste and the site waste minimization strategy topics.

Waste generation has increased at an alarming rate of 170 per cent in Australia from 1996 to 2016, while our population has risen by only 28 per cent in that period, with 40 per cent of total landfill coming from construction and demolition waste (Ritchie 2016). However, in Australian states, with stringent regulations on site waste recycling, such as New South Wales, landfill rates have been levelling off since 2005. There is a public acceptance that waste recycling benefits the broader community, with 51 per cent of Australian household waste being recycled, compared to 42 per cent, on average, in the European Union (Planet Ark 2018). Furthermore, this research has shown a preoccupation of engineering construction industry participants in this survey with site waste recycling; this is borne out by a claim that construction waste recycling/re-use recovery rates in one Australian state, South Australia, is as high as 80 per cent (Zero Waste South Australia 2013).

The overall Australian state and territory average construction waste recovery rate is 58 per cent (Hickey 2015). The 2016 Australian government, "*State of the Environment Overview*", tabled to parliament in March 2017, was a follow-on report from the previous 2011 document (SOE 2016). This report indicated that national recycling rates had improved over the past several years, however, a detailed review of the data showed that this 2016 report used recycling data from states/territories that were as old as 2010/11 and was collated into these updated findings (SOE 2016). This textbook's literature review for waste management in Australia has shown that the reliability of these government statistics has been challenged in several professional and trade publications and it is concluded that efforts should be made at the Federal Government level to determine current construction waste data more accurately on an annual basis as a benchmarking tool to assist with construction waste management regulatory decision-making.

Accordingly, it is concluded that a cultural shift can be successfully introduced through a national campaign, as was successfully implement with recycling in Australia. This change management process must be supervised by the government and include statutory regulation control measures and the embracement of this change in business culture by leaders of the engineering construction industry to filter through to all their employees and the public at large. The message for change being communicated should be that front-end site waste reduction presents as a critical activity to reduce project costs; enhance the environment by conserving non-replaceable material resources; reduce landfill footprint and carbon emissions; and provide social benefits due to more cost-effective public works programs and lowering residential building costs.

9.3 Front-End Site Waste Minimization Strategies Conclusions

The following front-end site waste minimization strategies are summarized in descending order, ranked using respondent survey data synthesis, using the abridged Yates (2013) moduli

I. Address waste generation reduction during project pre-planning to utilize designs that minimize waste using any of the following techniques: precast; prefabrication; pre-assembly or modularization.

II. Regarding the use of temporary construction materials, consider waste minimization processes as an example for concrete construction. Designers should specify concrete elements of similar dimensions, where practical. The use of steel shutters on repetitive formwork is recommended. It is important to consider whether the formwork is adequately treated and robustly fabricated, to allow re-use and whether orders are "just in time", to reduce material losses on-site.

III. Adopt a structured approach both for engineering design and in the determination of construction methodologies that involve waste minimization strategies.

IV. Innovative designs, construction components or construction processes should be integrated into construction projects to reduce site generated waste.

V. Firm's should use techniques that improve resource efficiency, equipment efficiency, material resource efficiency and allow for the training of manual labour.

VI. The contractors/builders, consultants and vendors should constructively work with clients to minimize change orders that make pre-ordered products redundant and suitable only for waste.

VII. Builders, contractors and engineering consultants should ensure a minimum amount of permanent and temporary materials are expended in the effective provision of client conforming construction/building products.

VIII. The contractor/builder and/or client should have a mandatory waste minimization plan developed as part of the project execution plan.

The thematic analysis of respondent data carried out on Questions 26–33, inclusive, of the survey questionnaire covering front-end site waste minimization strategies found that only 37.5 per cent of the 49 survey participants answered in the affirmative when asked if their company had a mandatory site waste minimization plan as part of the company's standard project execution plan. This was the lowest ranked strategy for this survey. This result is consistent with Udawatta et al. (2015), in which Australian project managers rated a company waste management plan as only the tenth most important factor in reducing site waste. Back in 1998, Faniran and Caban commented on the low number of respondents (58.1 per cent) in their landmark Australian site waste study who indicated that their company had a company waste management plan. Yates (2013) notes in their findings, funded by the USA's peak construction contracting body that only 50 per cent of the senior executives interviewed in that study on site waste noted that their company had a company site waste plan.

Pursuant to the above, it is concluded that the lack of a company site waste minimization plan is a major strategic impediment to the reduction of site waste at the front-end of the construction materials logistics chain. Accordingly, there must be a national agenda developed in Australia to ensure that a site waste management plan is a mandatory requirement for Australian engineering construction industry stakeholders, including federal, state and territory governments; capital works clients; engineering consultants; construction contractors; builders; vendors; and suppliers. This site waste management plan would contain, as a major component, strategies to reduce front-end site waste.

In addition, since knowledge transfer within the construction sector is difficult to achieve, a company site waste management policy would greatly assist in the dissemination of knowledge within a company (Argote et al. 2000). A standard site waste policy, company-wide, would also assist in a portfolio approach for a company executing its capital works and would greatly assist in the strategic management of capital assets (Eisenhardt and Zbaracki 1992).

Faniran and Caban (1998), Lingard et al. (1997) and Yates (2013), all state the case for the necessity for a company to have a robust site waste management plan in place to minimize construction waste. Osmani et al. (2008), Udawatta et al. (2015) and Yuan (2013), cogently argue that only a regulatory framework will ensure that site waste management processes are successfully adopted by the engineering construction industry.

An Australian code of practice for construction site waste minimization could be developed by the Australian Federal Government for mandatory national use, from which engineering construction industry consultants; contractors; builders; suppliers; vendors; and public/private sector clients would have to adopt. This national code of practice would be included in all capital works contracts. It would be the ultimate responsibility of the client project stakeholder to ensure this document's adoption and compliance, with a company site waste management plan from all of the contracted entities being a condition of tender submission. A project-specific site waste management plan could be developed by the prime contractor and approved by the client. Each site waste management plan from the proposed national site waste code of practice, to the client document down to the design consultant, contractor and supplier documents, along with the site waste management plan would all be complementary.

The apportionment of project risks, including construction waste management, resides with the client in equitable agreement with the contracted parties. However, it is concluded that it should be made abundantly clear by regulation that although the ultimate responsibility for project site waste resides with the client, all the contracted parties, contractors/builders, design consultants and suppliers, are jointly and separably responsible for site waste management for their particular scope of works.

These research findings lead the author to conclude that there are no new sources of waste and no new front-end site waste strategies. This is predicated on the survey participant responses for this research when requested to describe any other site waste sources and site waste reduction strategies in the survey questionnaire. Though participant comments were at all times valuable to the research, responses on "other" sources of waste and "other" waste reduction strategies were variations on sources and strategies nominated in the survey questionnaire for review. The research conclusions from the Faniran and Caban (1998) seminal work on-site waste are not very different from the findings in this research.

Triangulation of available data from other academic studies, such as Nagapan et al. (2012); Sugiharto et al. (2002); Udawatta et al. (2015); and Yates (2013), all indicate a degree of compatibility with the findings of this research even when accounting for the different sample sizes and the configurations and methods of data synthesis between some papers. The academic findings on front-end site waste minimization, despite offering numerous commercial advantages to client and contractor, as well as the greater community and the environment, are not being taken up by the engineering construction industry, and hence, a regulatory solution remains the only viable approach. However, the front-end site waste minimization solution has the potential to deliver major net cost savings to both the client and contractor. This is in marked contrast to the current regulatory drive in Australia to encourage recycling/re-use recovery processes by charging punitively high landfill disposal imposts on construction waste. With the exception of the 25 per cent of the waste levy allocation to waste management improvements by the Western Australian government, these waste management levy monies go to the consolidated revenue of all other state and territory governments (Ritchie 2016).

"Management" was the predominant theme in the thematic analysis of front-end site waste reduction strategies, synthesized from respondent comments provided by the survey questionnaire for this research. Only the top management cadre in government and in the private sector are capable of undertaking the enormous work culture change process that should be made to the Australian engineering construction industry to enable it to embrace front-end site waste reduction strategies. The major sub-theme for front-end site waste minimization strategies derived from the respondent survey comments was "FEED". This "design" theme and "FEED" sub-theme have been validated by a survey of participant rankings for the "yes/no/don't know" Questions 26–34, inclusive. The top three respondent replies in descending order were all design related, with the first two questions being FEED related, as follows:

I. Address waste generation reduction during project pre-planning to utilize designs that minimize waste using any of the following techniques: precast; prefabrication; pre-assembly or modularization.

II. Regarding the use of temporary construction materials, consider waste minimization processes as an example for concrete construction. Designers should specify concrete elements of similar dimensions, where practical. Steel shutters should be used on repetitive formwork. It should be considered whether the formwork is adequately treated and robustly fabricated to allow re-use and whether orders are "just in time" to reduce material losses on-site.

III.   Adopt a structured approach both for the engineering design and in the deter-
        mination of construction methodologies that involve waste minimization strate-
        gies.

   The conclusion is that early engineering design from project inception accounting
for site waste reduction considerations, moving forward through the engineering
design process, is the second most important strategy in ensuring front-end site waste
minimization. All of the above three design and construction process-related items
could be incorporated into the national government code of practice for site waste
management, as well as following the strategy from the abridged Yates (2013) model
used in this research. The reader's attention is drawn to the work of the University
of Exeter on its 10 point criteria to reduce site waste, mostly by front-end initiatives,
which may result in a two per cent reduction in total project costs (Vowles 2011).
Full details are available in Sect. 8.6, "Front-End Site Waste Minimization Strategies
Discussion".

9.4   Variances in Definition of Project Waste Scope Conclusion

This textbook examines the sources of physical waste and the strategies to reduce
front-end physical waste. Physical construction waste is defined as waste which arises
from construction, renovation and demolition activities, including land excavation or
formation, and civil or building construction with the waste product comprising of
physical inert temporary and/or construction material such as rebar, concrete debris,
timber offcuts and packaging waste (Alarcon 1997; Ferguson 1995; Formoso et al.
1999; Yates 2013). Some academic studies also account for non-physical waste in
their site waste modelling, which considers waste related to time and cost overruns for
a construction project, due to labour, material and equipment resource productivity
inefficiencies (Nagapan et al. 2012).
   The Australian Construction Industry Forum (ACIF) definition of project waste
suggests that the amount of waste on a project typically exceeds 30 per cent (Barda
2013). The ACIF further posits that a reduction of waste by only a third would
result in savings of A$10 billion per annum to the Australian construction industry
(Barda 2013). Josephson and Saukoriipi's (2007) definition of project waste argues
that the cost of waste adds 30–35 per cent onto the project capital cost. These two
definitions go far beyond the definition of "physical construction waste" adopted for
this research. For detailed definitions of project waste of the above two ACIF and
Josephson and Saukoriipi (2007) papers, refer to Sect. 8.8, "Variances in Literature
on Site Waste Definition Discussion".
   Physical construction waste amounts for around 13 per cent, on average, of the
project contract sum on building projects in Scotland (Zero Waste Scotland 2017).
Depending on the type of project, this research concludes that physical construction
waste in Australia adds an impost of an average of 6.5 per cent to the project cost.

The Master Builders Association states that physical site waste costs add an extra 10 per cent onto their projects (Doab 2018; MBAV 2004). Pursuant to the above, it is concluded that considerable caution should be taken when validating and correlating construction waste quantities and the cost of construction waste to projects across the available literature because project waste definitions, perspectives, viewpoints, scope and research methods vary markedly (Josephson and Saukoriipi 2007).

9.5   Validity of Waste Data from Australian Federal, State and Government Agencies and Other Regulatory Issues Conclusion

A review of the available Australian academic, professional and trade waste management literature, as outlined in this textbook via progressively more detailed literature reviews on construction waste minimization, indicates that there are a number of verifiable concerns surrounding Australian federal, state and territory governments' handling of waste management matters. One compelling example, Ritchie (2017), cites an Australian Federal Government document entitled "*State of the Environment Overview*", which was dated 2016 and tabled in parliament in March 2017. This document had data on waste management embedded with five-year-old data, which went back to 2010/2011, with inconsistent date baselines from each of the various states and territories (SOE 2016). Refer to Figure 2.1, "Summary of Data Used in the Commonwealth Government's 2016 "*State of the Environment Overview*" Report."

Hickey (2015) has strong views on the politicization by governments in Australia of waste management and believes that there have been instances of government agencies rejecting their own consultant findings because they do not align with current government construction waste control policies.

The author has found a lack of solid data on construction waste in Australia and, indeed, internationally. This type of data, such as the composition of material and approximate quantities of materials found in construction and demolition bulk waste, is information necessary for use by the public and private sectors to make informed decisions on waste control and, more particularly, construction waste management.

With the mooted possible effective collapse of the recycling industry in Australia, it is also concluded that it is now appropriate that governments in Australia be challenged to reassess their waste control strategies (Lasker 2018).

References

Alarcon, L. F. (1997). *Lean construction processes*. Chile: Catholic University Press.
Argote, L., Ingram, P., Levine, J., & Moreland, R. (2000). Knowledge transfer in organizations. *Organizational Behaviour and Human Decision Processes, 82*(1), 1–8.

Australian Construction Industry Forum (ACIF). (2017). *Latest Forecast*, viewed March to April 2017, https://www.acif.com.au/forecasts/summary.

Barda. (2013). *Stop the waste—30% is too much*, viewed March 2018 https://www.acif.com.au/news/opinion/stop-the-waste—30-is-too-much.

Chandler, D. (2016). Construction waste is big business. *Construction News,* 12 October, viewed May 2017–May 2018, https://sourceable.net/construction-waste-is-big-business/.

Core Logic. (2016). Cordell cost guide. https://www.corelogic.com.au/products/cordellcostguides. viewed October–December, 2017.

Doab, M. (2018). Average Cost of Construction Site Waste on Major Australian Projects. *Doab Estimation Enterprises Proprietary.* D'Est Estimating Model.

Eisenhardt, K. M., & Zbaracki, M. J. (1992). Strategic decision-making. *Strategic Management Journal, 13,* 17–37.

Faniran, O. O., & Caban, G. (1998). Minimizing waste on project construction sites. *Engineering Construction and Architectural Management, 17*(1), 57–72.

Ferguson, J. (1995). *Managing and minimizing construction waste: A practical guide.* London, UK: Institute of Civil Engineers.

Formoso, C. T., Isatto, E. L., & Hirota, E. (1999, July). *Method for waste control in the building industry.* Paper presented to the 7th Annual Conference of the International Group for Lean Construction, University of California, Berkley.

Hickey, J. (2015). *Landfilling construction waste in Australia*, viewed April to December 2017, https://sourceable.net/landfilling-construction-waste-in-australia/.

Hyder Consulting. (2011). *Department of sustainability, environment, water, population and communities—waste and recycling in Australia 2011 incorporating a revised method for compiling waste and recycling data.* Australian Commonwealth Government, viewed May to July 2017, http://www.wmaa.asn.au/event-documents/2012skm/swp/m2/1.Waste-and-Recycling-in-Australia-Hyder-2011.pdf.

Jain, M. (2012). Economic aspects of construction waste materials in terms of cost savings—A case of Indian construction industry. *International Journal of Scientific Research Publications, 2*(10), 1–7.

Josephson, P., & Saukoriipi, L. (2007). *Waste in construction projects: Call for new approach.* Sweden: Chalmers University of Technology Publication.

Lasker, P. (2018). Australia needs to start recycling and re-using its own waste, says industry struggling under China Ban. *ABC News,* 15 April, viewed 16 April 2018, http://www.abc.net.au/news/2018-04-15/australia-tossing-up-circular-approach-to-its-waste/9657342.

Lingard, H., Graham, P., & Smithers, G. (1997). Waste management in the Australian construction industry. In *13th Annual Kings College Cambridge UK ARCOM Conference,* Cambridge University.

Master Builders Association of Victoria (MBAV). (2004). *The resource efficient builder–A simple guide to reducing waste.* MBAV Guideline Publication.

Nagapan, S., Rahman, I., & Asmi, A. (2012). Factors contributing to physical and non-physical waste generation in construction. *International Journal of Advances in Applied Sciences, 1*(1), 1–10.

Osmani, M., Glass, J., & Price, A. (2008). Architects' perspectives on construction waste reduction by design. *Waste Management Journal, 28,* 1147–1158.

Planet Ark. (2018). *How does Australia compare to the rest of the world?,* viewed 25 April 2018, https://recyclingweek.planetark.org/recycling-info/theworld.cfm.

Ritchie, M. (2016). State of waste 2016—current and future Australian trends. *MRA Consulting Newsletter,* 20 April, viewed March to April 2018, https://blog.mraconsulting.com.au/2016/04/20/state-of-waste-2016-current-and-future-australian-trends/.

Ritchie, M. (2017). State of waste data (2017). *MRA Consulting Newsletter*, viewed March to April 2018. https://blog.mraconsulting.com.au/2017/03/29/the-state-of-the-waste-data/.

State of the Environment Overview (SOE). (2016). *2016 State of the Environment Overview*, viewed May 2017, https://www.environment.gov.au/science/soe.

Sugiharto, A., Hampson, K., & Sherid, M. (2002). Non value adding activities in Australian construction projects. Paper presented to the *International Conference for Advancement in Design, Construction, Construction Management and Maintenance of Building Products,* Griffith University, Australia, 2002.

Udawatta, A., Zuo, J., Chiveralls, K., & Zillante, G. (2015). Improving waste management in construction projects: An Australian study. *Resources, Conservation and Recycling, 101,* 73–83.

Vowles, F. (2011). *Construction strategy,* University of Exeter, viewed November to December 2017, https://www.exeter.ac.uk/media/universityofexeter/campusservices/sustainability/pdf/construction-strategy.pdf.

Waste Management World. (2016). *State of the Nation,* Waste Management World periodical, viewed November 2017 to February 2018, https://www.waste-management-world.com/au/report-state-of-waste-2016-current-and-future-trends.

Yates, J. (2013). Sustainable methods for waste minimisation in construction. *Construction Innovation, 13*(3), 281–301.

Yuan, H. (2013). Critical management measures contributing to construction waste management: Evidence from construction projects in China. *Project Management Journal, 44*(4), 101–112.

Zero Waste Scotland. (2017). *The cost of construction waste.* Government of Scotland, viewed February to June 2017, http://www.resourceefficientscotland.com/content/cost-construction-waste.

Zero Waste South Australia. (2013). Annual Report 2012–2013. *Government of South Australia,* viewed May to June 2017, http://www.zerowaste.sa.gov.au/upload/resource-centre/publications/corporate/3ZWSA%20Annual%20report%202013%20DE_02.pdf.

Chapter 10
Recommendations

10.1 Recommendations for PhD Higher Academic Research

10.1.1 Optimal Hours of Work on Australian Construction Sites

Out of the eight options that the author reviewed for further postgraduate research, the second highest ranked option was Optimal Work Duration on Site. The independent peer review panel who evaluated and recommended on the option selection believed that this was the other option that was worthy of research to develop an optimization model that provided optimal shift times and work weeks, with a number of variables such as site location; ambient conditions; and regulatory occupational health and safety constraints.

Refer to "long-list" Sect. 2.2.4, "Optimal Hours of Work on Australian Construction Sites", and "short-list" Sect. 3.2.1, "Option A "Short-List"—Optimal Work Duration on Site", along with peer review panel Sect. 3.3.6.2, "Workshop Panel Question No. 2", (on Option A—Optimal Work Duration on Site).

10.1.2 Further PhD Higher Research on Construction Waste Minimization

There are a number of areas of research that require investigation at PhD level. A potential PhD candidate would need to work closely with their proposed supervisors to ensure that a combination of these topics would be achievable over a full-time period of three years.

© Springer Nature Switzerland AG 2019
P. G. Rundle et al., *Effective Front-End Strategies to Reduce Waste
on Construction Projects*, https://doi.org/10.1007/978-3-030-12399-4_10

10.1.2.1   Front-End Site Waste Minimization Cost–Benefit Analysis

It was the intention of the author to undertake an upper-level cost–benefit analysis (CBA), which would provide evidence of the potential major advantages of front-end site waste minimization on Australian construction projects. In fact, Faniran and Caban's (1998) seminal work on construction waste also recommended that a CBA in this area needed to be executed. The reader's attention is drawn to Sect. 3.3.6.3, "Workshop Panel Question No. 3", which is a response to a peer review panel workshop question that directed the author to carry out a further literature review on construction site waste reduction, in which it was recommended by the author that a CBA be prepared in this textbook. A detailed literature review was subsequently prepared on a proposed CBA to assist in the preparation of the research framework. During the confirmation phase of this book, it was the opinion of the assessment panel that a CBA, though valuable, was a topic for higher research and that the remaining scope for this research was suitable. The author discussed this recommendation with a nominated SCU professor of economics, who subsequently validated the assessors' findings in November 2017 in writing to the authors that the CBA would be so detailed for front-end site waste minimization, involving not only commercial benefits, but ecological factors such as carbon emission and degradation of the riverine system, along with social benefits to the community at large, that this work was a PhD level area of research. It is strongly recommended that a comprehensive cost–benefit model be prepared for construction waste minimization as a marketing tool to alert the government, the Australian engineering construction industry and the community of the pressing requirement to introduce these high-value initiatives using mostly known, simple and available strategies, at very limited capital expenditure. The academic research has shown that there is a void in the literature with respect to this proposed research and a global literature review has indicated that there is also a literature gap and a need for a comprehensive CBA model.

10.1.2.2   Front-End Site Waste Minimization Case Study

A case study is recommended that uses all the findings from this textbook on sources of waste and waste strategies as a real-world validation of this research, which could look towards benchmarking the commercial, environmental and social benefits as a result of the following these initiatives. A case study should be carefully chosen that would be of practical value to the Australian engineering construction industry, preferably with the support of a peak construction industry body that would encourage dissemination of the attained knowledge amongst consultants, builders/contractors, practitioners and suppliers. Most importantly, the developer and builder/contractor should be willing to accept the impost of academic intrusion on a commercial enterprise, even with the knowledge that the case study is aimed at commercially benefiting the project.

It is recommended that an urban high-rise building project would be an ideal case study since this project sector is currently driving construction in Australia, along with infrastructure development. A high-rise project case study would have international appeal for research findings, and a small horizontal footprint would assist in tracking the material waste flow. The author suggests that it would be appropriate if a environment client could be identified who would integrate the case study with the developer on a forthcoming project. Perhaps even a similar university capital works, large public building such as the recent SCU Gold Coast capital works program would prove suitable.

10.1.2.3 Best Practice Manual for Front-End Waste Minimization

There is local and international government literature available on waste management, of which a significant component is general in nature. The academic skill required will be in developing a serious document for practitioner use, which could be a template for the eventual development of an engineering standard in Australia as a mandatory code of practice. It is recommended that academia is a good starting point to develop this key document towards the regulatory approach necessary to lead the engineering construction industry and capital works clients towards a front-end waste reduction future, not fuelled by punitive levies, but rather by very tangible commercial, environmental and social benefits.

10.1.2.4 Site Waste Dynamic Model

It is recommended that quantitative research be undertaken to develop a dynamic model that simulates a project site's flow of construction waste for Australian projects. This topic was deemed by the authors to be beyond the scope of this book. Refer to Sect. 3.3.6.3, "Workshop Panel Question No. 3", on Option B, Construction Site Waste Reduction, for the full details.

A dynamic waste model could simulate the complicated construction site waste supply chain, commencing with off-site manufacture and fabrication of permanent and temporary materials and ending with construction material wastage being placed in landfill, buried in regulatory approved locations on site, or recycled as a new product for re-use. Dynamic modelling has many benefits for project stakeholders, including application for planning purposes to determine wastage type and volume predictions, which benefits contractors by planning and optimizing waste materials and assists the regulatory authority by aiding landfill footprint calculations. Once calibrated, dynamic modelling can also be used to check actual wastage progressively on-site versus model simulation to determine whether site material wastage is being adequately controlled (Hao et al. 2007, 2008, 2010).

10.1.2.5   Site Waste Optimization Model

This textbook recommends the development of an effective optimization model via PhD research that will assist in determining the optimal process to efficiently minimize the volume of construction waste on a project site (Yeomans 2004). The purpose of this optimization study should be to provide practitioners with a model that will assist in the planning of optimal methodologies for the collection, allocation and disposal of physical construction waste. This topic was deemed by the authors to be beyond the scope of a this book. Refer to Sect. 3.3.6.3, "Workshop Panel Question No. 3", for the full details.

10.1.2.6   Broad Definition of Project Waste—Minimization Research Study

A study is recommended to undertake PhD standard research that ties together all of the available fragmented literature on the definition of project waste and carries out research on all of the available opportunities. This could include a study of physical waste and non-physical waste, including site resource inefficiencies causing negative cost and schedule impacts, and the ways of reducing unproductive engineering design caused by engineering consultants and architects providing drawing documentation that will not be used in construction. This research could also examine constant drawing reiterations at the development stage (due to poor stakeholder coordination and unclear scope definition), the imposts to a project by a turgid planning and development approval regulatory process, and the hidden costs of material defects such as high insurance premiums.

10.1.2.7   SCU as a Postgraduate Research Engineering Sustainable Development Hub

The author is optimistic that Southern Cross University (SCU) has the capability to become an academic hub collaborating with other Australian universities, such as UNSW, RMIT and University of South Australia, that encourages postgraduate Masters of Science and PhD students to carry out higher research into sustainable development related to the engineering construction industry, which is the world's largest user of raw materials, providing consistently high employment and a litmus paper for economic health, from developing to developed countries. It is therefore recommended, in view of the current crisis in the Australian recycling waste industry, that SCU endeavours to build on the research work of this textbook by encouraging future science and engineering science postgraduate students to continue this work.

10.2 Recommendations for Further Research on Site Waste

10.2.1 Potential Sources of Site Waste Identification Model

A refinement to the Faniran and Caban (1998) sources of site waste identification model implemented by the author in this research was the separation of the question referring to "design changes being a source of site waste" into two separate questions, namely (i) "client-initiated design changes"; and (ii) "contractor-initiated design changes". This adaption to the Faniran and Caban (1998) model was made by the author in response to the literature review on site waste minimization strategies, in which it was noted that client-directed design changes presented as a significant cause of site waste (Yates 2013). As this was determined as the main source of site waste in the research, this revision to the model is recommended.

Another adaption to the Faniran and Caban (1998) sources of site waste identification model made by the author was the addition of one question to complete the survey on sources of site waste. This question requested that respondents provide comments as to any other potential sources of site waste; respondent comments provided a rich pool of data that allowed the author to undertake a thematic analysis. Therefore, this additional question is also recommended for inclusion, as a further positive adaption to the model and the thematic analysis approach is recommended for interpreting the model output for this particular question.

As per the Nagapan et al. (2012) findings, in which it was determined that "criminal activity and site accident sources of waste" were seen as minimal causes of site waste, which was confirmed in this textbook it is recommended that due prudence is used by future researchers in including the "criminal activity waste source" and the "site accident source" questions in the survey. The future researcher could determine via the project particular environment and location whether site waste created by vandalism and theft, or material waste through poor construction practices leading to high accident rates, would justify the inclusion of these two questions. The extensive global literature review of Nagapan et al. (2012) on-site waste indicates that this is generally not the case; this is borne out by local academic literature on-site waste (Sugiharto et al. 2002; Udawatta et al. 2015).

In conclusion, the author recommends that where, in the opinion of a researcher, that a potential for site waste is unlikely to be either criminal activity or via site accidents, these two questions are deleted as an adaption to the Faniran and Caban (1998) model. In any event, any such sources of waste would be picked up in the final question of the survey asking for other possible sources of waste. The author posits that the fewer survey questions asked, the better the quality of the participant responses, and the more likely that the survey questionnaire shall be fully completed; this method is supported by academic research (Creswell 2013).

One another refinement to the Faniran and Caban (1998) sources of site waste model, also adapted for use by Nagapan et al. (2012), has been the interpretation of respondent data. Both Faniran and Caban (1998) and Nagapan et al. (2012) only used a severity index to rank the sources of waste. Although the author followed this accepted methodology and calculated the severity index from the model, they also ranked all participant answers for each question to obtain a ranking taking all opinions into account (not just "strongly agree", to obtain a severity index). It is recommended that the synthesis of the model output by using both ranking methods provides a more complete view as to the respondents' overall assessment of the major sources of site waste.

10.2.2 Front-End Site Waste Minimization Model

The Yates (2013) site waste minimization strategies model, used for research purposes to investigate the status of construction waste reduction practices of the USA's top contracting and building companies, was abridged for use in this research. As this textbook was concerned with front-end site waste minimization, any of the Yates (2013) survey questions that solicited comments from respondents on recycling/re-use downstream waste recovery processes were excised from the model and the author deemed that this research uses an abridged, rather than an adapted, form of the model. Though the Yates (2013) paper cites data providing advice on the number of contracting firms who have work management processes in place, it was not clear whether this question was addressed through interview or via the survey model. The abridged front-end site waste minimization strategies model used in this research contained a question that solicited participant responses as to whether their firm had a mandatory front-end site waste plan as part of the company's processes and procedures. It is recommended that the Yates (2013) model includes such a question on a mandatory corporate waste plan.

Another recommended adaption to the Yates (2013) site waste minimization strategies model made by the author was the addition of a final question to complete the survey on front-end site waste minimization strategies. This question requested that respondents provide comments as to any other potential front-end strategies that can be used to reduce site waste. This question provided a rich pool of data via the respondent's comments, which validated that the model addressed all available waste minimization strategies. The reason for this was that respondent comments were versions of the strategies outlined in the model survey.

The final recommendation regarding the Yates (2013) model concerns the synthesis of model data output. The author diligently followed the Yates (2013) approach, recording all respondent comments for each survey question in tabular format, and ranking each participant response into a set of "yes", "no" or "don't know" percentages of the total respondent sample number, as a method of model output presentation to assist in interpreting the findings of this research. However, each set of respondent comments was also synthesized using the thematic analysis approach and the author was consequently rewarded with a window into an otherwise unseen set of themes and sub-themes that interwove impacting factors:

 I. Between themes and sub-themes;
 II. Within each theme and sub-theme, as per each individual sub-theme question;
III. Of common threads in the themes and sub-themes, recurrent in all of the survey questions for sources of site waste and front-end site waste minimization topics; and
IV. Importantly, cross-linkages of themes and sub-themes between both the sources of site waste topic and the front-end site waste minimization strategies topic.

Pursuant to the above, the author recommends the use of the thematic analysis approach as a powerful qualitative research method to provide deep insight into site waste minimization strategies using the Yates (2013) site waste minimization strategies model.

10.2.3 Further Research on Poor Weather as a Potential Source of Waste

It is recommended that further research by qualitative survey be undertaken to determine the one inconsistency, which was noted during the triangulation of synthesized research data findings for this study, with three other construction waste studies (Faniran and Caban 1998; Nagapan et al. 2012; Sugiharto et al. 2002), two of which were completed in Australia. Faniran and Caban (1998) carried out seminal work on construction waste in Australia (an adapted waste source identification model from this research was used for survey data synthesis in this book) and Nagapan et al. (2012) carried out a Malaysian study on construction waste. Although Sugiharto et al. (2002) used a different approach methodology, all three academic papers rated poor weather as a high source of potential waste from engineering construction industry surveys. However, this textbook rated the participant concerns in this survey as being minimal. The size of the respondent data sample used for this research was considerably larger than the other three studies used for validity purposes. The other potential sources of waste findings were very consistent when compared between this textbook and the other researched papers. Refer to Table 7.4, "Comparison of Waste Sources Research Results and Three Other Papers", Sect. 7.2.2, "Comparison of Part 2 Survey Data with Other Academic Journals for Sources of Site Waste".

10.2.4 Interview Independent Peer Review Panel

It is recommended that ten of the eminent industry peers who reviewed and provided recommendations on the selection of topic "short-list" and ultimately, the "Go Forward" case for detailed academic research, be interviewed to gauge their attitude and comments regarding the conclusions and recommendations of this textbook.

10.2.5 Establish a Research Committee to Reduce Packaging Waste Reduction

Packaging and pallet waste of permanent and non-permanent construction materials was such a continual topic of respondent comments in this research that an extra survey question was circulated to a discrete sample pool of 50 construction engineering specialists from the original survey pool of 102 people. Seventeen participants responded, with only two respondents, both clients, indicating that provisions were made with suppliers to reduce packaging waste as a condition of the shipment of physical material to site; the balance of responses was 12 "no" allowance made for packaging waste reduction and three "don't know" responses.

Excessive product packaging material waste is also a current topic on the national agenda caused, in no small part, by the Chinese ban on the export of Australian recycled waste to China since the start of 2018 (Lasker 2018). It is strongly recommended that a small working committee be formed, with representatives from government, peak engineering construction practitioner bodies and academia, to interface with major construction vendors and suppliers to develop a strategy for the reduction of construction material packaging and associated materials, such as materials handling products, for example pallets, bins and cages. This recommendation to find strategies to reduce this waste source could just as readily have been inserted in the next Sect. 10.3.

10.3 Recommendations for Government Regulatory Initiatives

10.3.1 Improve Efficacy, Frequency and Scope of Site Waste Data

There is a degree of scepticism in Australian academic, professional and waste management industry literature regarding the efficacy of Australian federal, state and territory government statistics on all forms of waste in Australia and that waste management in Australia has become politicized by successive governments (Chandler 2016; Hickey 2015). Ritchie (2017) makes the point that the 2016 "State of the Environment Overview" document tabled in parliament in March 2017, contained five-year-old waste management statistics for some states and territories, which went back as far as 2010/2011 (Ritchie 2017; SOE 2016). The previous Federal Government "State of the Environment Overview" document was produced back in 2011; this data needs to be produced annually.

Pursuant to the above, it is recommended that the Australian Federal Government look towards the provision of accurate and timely waste management statistics, with a single benchmark date for state and territory data input, to assist interested stakeholders in making informed decisions. Since the Australian construction industry accounts for approximately 40 per cent of all waste produced in Australia and a third of all solid landfill, government regulatory factors have a significant impact on this industry, making it the largest waste producing industry in Australia (Hyder Consulting 2011; Waste Management World 2016).

During the course of this research, the author has discovered that there is a lack of practical statistical data on the composition and quantity of C&D waste in both Australian and international literature. The composition and quantity of construction waste from each primary sector of the Australia construction industry, ranging from residential tract housing to heavy civil infrastructure construction, also need to be inventoried on a regular basis.

The provision of this data may assist with the development and implementation of environmental legislation and this current and reliable information shall prove valuable to both the waste management industry and the engineering construction industry. This is a critical recommendation, if serious inroads are to be made on front-end construction waste reduction in Australia.

10.3.2 Australian Government to Develop a National Landfill Asset Register

Australian Federal Government data on landfill sites notes that there is a minimum of approximately 600 medium to large facilities currently in operation, and while the total number of landfill facilities are unknown, there could be as many as 2000 unregistered and unregulated landfill sites in Australia (Waste Management World 2016). The Australian construction industry is a major user of private landfill sites and any reduction in solid waste disposal by front-end waste construction minimization may lead to a reduction in these often unregistered landfill sites, many of which are unlined and hence a cause for environmental concern due to possible local contamination factors (Waste Management World 2016).

It is recommended that the Australian Federal Government expediently establishes an asset register of all registered and unregistered landfill sites in Australia to allow the collaborative and proactive development of a national landfill management plan. In addition to environmental and community benefits, this plan shall provide the engineering construction industry with a blueprint for compliant and economical solid waste disposal.

10.3.3   Government Waste Management Strategy Review

It is recommended that all nine Australian governments now proceed to challenge their downstream dominated recycling strategies (which use European Union waste recovery metrics to determine national performance benchmarks) by taking a basic risk management view of waste control. This view dictates that the removal of a potential risk (front-end waste reduction) is far more efficient and economical than to mitigate that risk (downstream recycling of waste). Figure 10.1, "Hierarchy of Waste Management", captures this strategy for construction waste minimization. The re-evaluation of this country's waste management platform is urgently required, in no small part due to two external drivers of change, namely the Chinese ban on the importation of Australian recycled waste material and the worldwide 30 per cent collapse in recycled product commodity prices (Lasker 2018). Additionally, the exigencies of Australia, due to its enormous land mass and very small population, have broad sweeping implications for any correlation with European Union countries when evaluating construction site waste management and importantly, waste management in general.

There shall always be an important place for construction waste re-use/recycling recovery practices; however, these strategies need to be economically viable for the contractor, government and waste management stakeholders. The punitively high landfill levy on solid waste in New South Wales has commercially driven the behaviour of construction contractors from both New South Wales and Victoria to dispose of solid waste across the border, to south-east Queensland landfill sites with marginal levy rates. This has resulted in negative impacts, such as a large increase in truck movements along the Pacific Highway and the loss of revenue for the Victorian and New South Wales governments. Only the Western Australian government spends any of its landfill levies on waste management (25 per cent) while all other states return levy costs to consolidated revenue.

There shall always be a requirement for a landfill levy based on the market viability of recycling. In 2013, the Queensland Government bought in a A\$35 per tonne industry waste levy, which created an immediate increase of recycling rates. However, when the levy was removed, 18 months later with a government change, recycling rates plummeted by 15 per cent. It is recommended that a national landfill levy be struck for all states and territories, based on waste management process imperatives, established via economic analysis, as an outcome of the suggested government strategic review of waste management.

The author is hopeful that the recommended Australian Federal Government strategic review of waste management shall bring front-end waste strategies to their rightful place at the forefront of the hierarchy of waste control in Australia and that a program be developed to effect changes in current waste practices in Australia. Since the engineering construction industry is the largest industrial generator of both landfill solid waste and overall waste in Australia, regulatory management encouraging a move towards front-end construction waste strategies will be a key plank in the revised strategic platform.

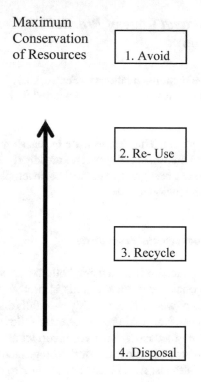

Maximum
Conservation
of Resources

1. Avoid

2. Re- Use

3. Recycle

4. Disposal

**Hierarchy of Waste
Management:**

(1) Avoiding waste

(2) Re-using materials

(3) Recycling and reprocessing materials

(4) Waste disposal (if first three options are not possible)

Fig. 10.1 Hierarchy of waste management. Adapted from: Faniran and Caban (1998)

10.3.4   National Code of Practice for Construction Waste Control

In line with the recommendations in Sect. 10.3.3, "Government Waste Management Strategy Review", it is further recommended that a *National Code of Practice for Construction Waste Management*" is prepared by the Australian Federal Government and that this document becomes a regulatory requirement on all capital works projects in Australia. This research study has shown that the sources of waste and front-end waste minimization strategies are largely known. See Sect. 10.1.2.3, which recommends that a best practice manual be developed by academia as a starting point for developing a national code.

10.3.4.1   Procurement Procedures

Synthesis of this research data, using both the severity index and thematic analysis, has identified quality control shortfalls in procurement procedures as the cause of several sources of waste. These quality shortfalls entailed either missing project procurement processes or a failure to correctly follow project procurement processes. Many of the identified issues involved incorrect material take-offs, incorrect materials ordering, inadequate site warehouse storage facilities and poor on-site handling of construction materials. All of these issues point to the damage to physical materials delivered to sites, which are then rejected and go to construction waste. It is recommended that the national code of practice for construction waste management also incorporates a "procurement plan" that incorporates these interdependent processes.

The planning and control of site material flow were another major concern identified in the Part 2 sources of waste survey for this research. These concerns were reiterated in Part 3, front-end site waste minimization strategies, via thematic analysis. The thematic analysis indicated that there was strong interest in the adoption of an integrated project material logistics supply chain by survey participants. This site materials supply chain would tie in and coordinate all stakeholders, including clients; design consultants; suppliers; vendors; contractors/builders and regulatory authorities. This integrated project material logistics supply chain would be in place from project inception through the life of the project and include the material warranty period. It is recommended that this procurement plan should also incorporate the requirement for an integrated project material logistics chain.

10.3.4.2   Design Plan

This research also showed that there was a transparent nexus between the importance of design for both the sources of site waste study and the front-end site minimization strategies study components of this textbook. The two highest potential waste sources for this research were both design related (design changes and design/detailing errors). Data interpretation via thematic analysis (from respondent surveys on front-end site waste minimization strategies) clearly showed that the majority of the survey participant remarks on this exhaustive analysis focused on the importance of early front-end engineering design (FEED) as a means of designing out potential sources of waste. This "FEED" sub-theme was a consistent factor in economical construction waste reduction strategies in the literature review. The literature review also uncovered that design changes in general and design errors were two reoccurring sources of waste in other academic research.

Accordingly, it is recommended that a "design plan" be incorporated into the suggested national code of practice for construction waste management, to ensure that proper attention is paid, particularly in the early phases of the design process, to site waste reduction measures.

References

Chandler, D. (2016). Construction waste is big business. *Construction News,* 12 October, viewed May 2017–May 2018. https://sourceable.net/construction-waste-is-big-business/.

Creswell, J. W. (Ed.). (2013). *Qualitative inquiry and research design—choosing among five approaches.* USA: Sage Publications.

Faniran, O. O., & Caban, G. (1998). Minimizing waste on project construction sites. *Engineering Construction and Architectural Management, 17*(1), 57–72.

Hao, J. L., Hills, M. J., & Huang, T. (2007). A simulation model using system dynamic method for construction and demolition waste management in Hong Kong. *Construction Innovation, 7*(1), 7–21.

Hao, J. L., Hill, M. J., & Shen, L. Y. (2008). Managing construction waste on-site through system dynamics modelling: The case of Hong Kong. *Engineering, Construction and Architectural Management, 15*(2), 103–113.

Hao, J. L., Tam, V. W. Y., Yuan, H. P., Wang, J. Y., & Li, J. R. (2010). Dynamic modeling and demolition waste management processes: An empirical study in Shenzhen, China. *Engineering, Construction and Architectural Management, 17*(5), 476–492.

Hickey, J. (2015). *Landfilling construction waste in Australia,* viewed April to December 2017. https://sourceable.net/landfilling-construction-waste-in-australia/.

Hyder Consulting. (2011). *Department of sustainability, environment, water, population and communities—waste and recycling in Australia 2011 incorporating a revised method for compiling waste and recycling data,* Australian Commonwealth Government, viewed May to July 2017. http://www.wmaa.asn.au/event-documents/2012skm/swp/m2/1.Waste-and-Recycling-in-Australia-Hyder-2011.pdf.

Lasker, P. (2018). Australia needs to start recycling and re-using its own waste, says industry struggling under China Ban. *ABC News,* 15 April, viewed 16 April 2018. http://www.abc.net.au/news/2018-04-15/australia-tossing-up-circular-approach-to-its-waste/9657342.

Nagapan, S., Rahman, I., & Asmi, A. (2012). Factors contributing to physical and non-physical waste generation in construction. *International Journal of Advances in Applied Sciences, 1*(1), 1–10.

Ritchie, M. (2017). State of waste data (2017). *MRA Consulting Newsletter,* viewed March to April 2018. https://blog.mraconsulting.com.au/2017/03/29/the-state-of-the-waste-data/.

State of the Environment Overview (SOE). (2016). *2016 State of the Environment Overview,* viewed May 2017. https://www.environment.gov.au/science/soe.

Sugiharto, A, Hampson, K & Sherid, M. (2002). *Non value adding activities in Australian construction projects.* Paper presented to the International Conference for Advancement in Design, Construction, Construction Management and Maintenance of Building Products, Griffith University, Australia.

Udawatta, A., Zuo, J., Chiveralls, K., & Zillante, G. (2015). Improving waste management in construction projects: An Australian study. *Resources, Conservation and Recycling, 101,* 73–83.

Waste Management World. (2016). *State of the Nation,* Waste Management World periodical, viewed November 2017 to February 2018. https://www.waste-management-world.com/au/report-state-of-waste-2016-current-and-future-trends.

Yates, J. (2013). Sustainable methods for waste minimisation in construction. *Construction Innovation, 13*(3), 281–301.

Yeomans, J. (2004). Improved policies for solid waste management in the municipality of Hamilton-Wentworth, Ontario. *Canadian Journal of Administrative Sciences, 21*(4), 376–382.

Chapter 11
Research Contribution

This chapter shall briefly summarize the contribution that this research has made to academia, practice and the broader community. This section uses Chap. 4, "Core Study Objectives", produced at the mid-point of this research for the SCU confirmation tollgate approval process, as a template to evaluate these results.

11.1 Achievement of Goals

Refer to Sect. 3.3.5, ""Short-List" Decision Analysis to Determine "Go Forward" Case(s)", which provides that the independent peer review panel approved that *"Front-End Strategies to Reduce Waste on Australian Construction Projects"* complies with all six of the KPIs for this textbook. Refer to Sect. 8.3, "Research Question Discussion", which provides an explanation as to why this book has met the research question and the sub-research question criteria. Refer to Appendix C.4, "Void in the Australian Literature", which provides justification for the contribution of this textbook to filling part of the gap in the literature.

The knowledge transfer of this work is via publication of this peer-reviewed textbook. A journal article will be written from this book, and the author was a keynote speaker in Indonesia at a management seminar on the 16 May 2018; the author's speech, on sustainable management, covered this research topic in part. The Master Builders Association (New South Wales) has requested that the author provides a speech on this topic, post-publication.

See Sect. 9.1, "Preamble to Key Conclusions", for a comprehensive assessment of the triple bottom line commercial, ecological and social benefits provided by the implementation of this textbook's initiatives.

© Springer Nature Switzerland AG 2019
P. G. Rundle et al., *Effective Front-End Strategies to Reduce Waste on Construction Projects*, https://doi.org/10.1007/978-3-030-12399-4_11

11.2   Contribution to Australian Engineering Construction Industry

A primary aim of this research has been to produce a study that has value for practitioners for general use and, ultimately, as a cost-saving initiative. From the inception of this research, improvement of the Australian engineering construction industry was a primary consideration as denoted by the title, "*Effective and Efficient Methodologies in the Australian Engineering Industry*", with the book's definition of efficiency being: "meeting all internal requirements for cost, margins, asset utilization and other related efficiency measures". This definition is clearly a contractor's metric. Further, the KPIs, developed as a tool to tollgate the eight options that were originally researched, have a strong emphasis on practical implementation.

These KPIs are noted as follows: (i) the research will be of transparent commercial benefit to contractor/engineering houses, client end-users and the community, at large; (ii) the proffered solutions can be readily and therefore expediently implemented; (iii) the topics selected for investigation shall provide maximum stakeholder benefits; (iv) the solutions to the identified inefficiencies and ineffective practices are, by and large, available within the academic and professional international body of knowledge; (v) the textbook must be practical in nature and address a void in the Australian engineering construction business, and the work must be valuable to this industry; and (vi) the identified research must broadly comply with a triple bottom line philosophy (Slaper and Hall 2011), and commercial, social and ecological benefits will be provided by these options.

Pragmatism was selected as the most appropriate interpretive framework to take the research forward using a qualitative methodology. The nature of this research, with a fundamental aim of benefitting engineering practitioners and the engineering construction industry, and an abiding interest in finding practical solutions to problems, means that a pragmatic approach was suitable for focusing on research outcomes that directly address the research question. One of the primary purposes of this research project has been to draw attention to the "Hierarchy of Waste Management" (see Fig. 10.1), which states that it is more efficient to prevent or minimize site waste at the front end of the logistics chain, instead of the more expensive and less environmentally sound downstream treatment via recycling. The collapse of recycling commodity prices, coupled with the ban on import by China of the 30 per cent per annum of recycled waste Australia has been shipping to that country, is threatening the commercial liability of recycling, with the Australian construction industry being a major recycling source (Lasker 2018). Punitively high landfill levies in some Australian states, particularly New South Wales, are a significant source of consolidated state revenue and is used as a mechanism to encourage contractors to recycle (Hickey 2015).

The main thrust of this book is to place an emphasis on front-end site waste reduction strategies, which can save contractors around two per cent of contract value (Vowles 2011). These minimization strategies have been identified, and they require no major capital to implement and no innovative, high-risk methods. The author argues that these strategies can be readily applied to the Australian engineering construction industry.

11.3 Regulatory Impact of This Textbook

This textbook has provided a number of recommendations on regulatory matters. The reader's attention is drawn to Sect. 10.3, "Recommendations for Government Regulatory Initiatives", for the full details.

11.4 Broader Community Benefits from This Textbook

A brief summary of the possible benefits to the community at large, as a result of this research, is noted as follows:

  I. Savings to the Australian construction industry of around A\$4.36 billion per annum due to front-end waste reduction, calculated in this textbook would benefit the community by saving funds on public works projects.
  II. Reduced construction costs on public housing and private residential buildings.
  III. It is likely that some of this A\$4.36 billion per annum savings would be reinvested into capital work, which would foster higher employment and create higher domestic consumption, and possibly government consumption, which in turn would raise gross domestic product (GDP).
  IV. Carbon emissions would decrease.
  V. Unproductive power, water and gas use would decrease.
  VI. Unproductive use of irreplaceable raw materials for construction material manufacture would decrease.
  VII. Less degradation of groundwater and riverine runoff systems, due to a decrease in landfill footprint.
  VIII. A reduction in unregistered and often unlined landfill sites.
  IX. A reduction in registered landfill sites.
  X. A reduction in heavy truck movements due to a decrease in recycling and less material transport.

11.5 Research Impact of This Textbook

Please refer to Sect. 10.1, "Recommendations for PhD Higher Academic Research", and Sect. 10.2, "Recommendations for Further Waste Minimization Research", for a detailed summation of the research impact that this textbook has made.

References

Hickey, J. (2015). *Landfilling construction waste in Australia*, viewed April to December 2017, https://sourceable.net/landfilling-construction-waste-in-australia/.

Lasker, P. (2018). Australia needs to start recycling and re-using its own waste, says industry struggling under China Ban. *ABC News*, 15 April, viewed 16 April 2018, http://www.abc.net.au/news/2018-04-15/australia-tossing-up-circular-approach-to-its-waste/9657342.

Slaper, T., & Hall, T. (2011). The triple bottom line: What is it and how does it work. *Indiana Business Review, 8*(1), 4–8.

Vowles, F. (2011). *Construction strategy*. University of Exeter, viewed November to December 2017, https://www.exeter.ac.uk/media/universityofexeter/campusservices/sustainability/pdf/construction-strategy.pdf.

Conflict of Interest

To the best of the author's knowledge, there has been no conflict of interest matters between the author and the party or parties in the execution of this academic work.

© Springer Nature Switzerland AG 2019
P. G. Rundle et al., *Effective Front-End Strategies to Reduce Waste on Construction Projects*, https://doi.org/10.1007/978-3-030-12399-4

Appendix A
Executive Summary

The Australian engineering construction industry is the largest non-services sector in this economy, accounting for 8.1 per cent of the gross domestic product (Gruszka 2017, p. 4). However, during 2014–15, the Australian Bureau of Statistics (ABS) ranked the "rate of innovation" in the Australian construction sector as the third lowest of all businesses (ABS 2015). Competitive advantage, gained by adopting effective and efficient engineering construction methodologies, is important within the Australian engineering sector. This is because many international engineering consultants and contractors enter the local market due to our stable political, social and economic foundations (Austrade 2018; Gruszka 2017).

The purpose of this textbook is to undertake a global literature review of capital works project practices and evaluate the eight most appropriate effective and efficient methodologies that will potentially add triple bottom line value to the Australian engineering construction industry.

After a two-step independent review process, using a panel of eminent industry and academic peers, a "Go Forward" case shall be selected for further detailed research. The "Go Forward" case is the option selected for further investigation in the next phase of an engineering study.

The preliminary eight options were chosen after an initial review of the literature, selected against key performance assessment criteria. This 'long-list' was comprised of the following options: (i) Knowledge Management; (ii) Lean Construction; (iii) Construction Contract Procurement Practices; (iv) Optimal Work Duration on Site; (v) Construction Site Waste; (vi) Rationalization of Australian Construction Safety Regulations; (vii) Sustainable Construction Labour Force; and (viii) Portfolio Project Development.

The key performance indicators (KPIs) included a requirement for options to be readily implemented; to be of transparent commercial benefit to client/consultant/contractor-builder stakeholders; and to offer a triple bottom line philosophy, which would provide financial, ecological and social benefits. The eight-option 'long-list' was appraised for Australian conditions by the peer review panel who selected a four-option 'short-list' for further research by the author (Peter

© Springer Nature Switzerland AG 2019   299
P. G. Rundle et al., *Effective Front-End Strategies to Reduce Waste
on Construction Projects*, https://doi.org/10.1007/978-3-030-12399-4

G. Rundle) of the available literature. These four methodologies were (a) Optimal Work Duration on Site; (b) Construction Site Waste Reduction; (c) Rationalization of Australian Construction Safety Regulations; and (d) Sustainable Construction Labour Force. The author then prepared four 'short-list' option briefs that were used by the second panel of eminent industry and academic peers, who selected the "Go Forward" option using the risk-based Kepner and Tregoe (2013) decision analysis process via an independently facilitated workshop. The peer review panel collaboratively assessed that the Construction Site Waste Reduction option was the most appropriate option and recommended this case for detailed research.

The subsequent extensive literature review on construction waste confirmed that there was a void in the academic writing, which identified the need to develop front-end strategies to reduce waste on Australian construction projects, as opposed to the recovery and recycling of construction site waste products at the end of the site waste supply chain.

The progression in the research topic rationale can be traced by the three research questions as this study unfolded. The initial research question was "*what effective and efficient methodologies are available to add triple bottom line value to the Australian engineering construction industry?*" A new research question was developed, after the selection of the "Go Forward" case, which superseded the initial research question. This revised primary research question was "*what effective and efficient front-end strategies are available to reduce waste on Australian construction projects?*".

Further literature research showed that, to the best of the author's knowledge, there has been no prior academic research undertaken in Australia that combines concurrent research into both front-end site waste reduction and potential sources of construction waste, while using qualitative thematic analysis and involving such a broad spectrum of engineering construction industry participants for the formation of the survey sample. Thematic analysis is a common form of qualitative analysis that involves highlighting, interrogating, examining and recording themes (that is, patterns) within the data. The adoption of the seminal Faniran and Caban (1998) sources of site waste model allowed comparisons to be carried out with similar local and international studies. Accordingly, a research sub-question was prepared to cover this study scope addition—"*what are the major sources of waste on Australian construction sites?*".

Next, the author's study objectives were outlined, including possible client/contractor savings to the Australian engineering construction industry. The several ecological benefits, such as a reduced national landfill footprint and a decrease in carbon emissions, were noted. Finally, broader community benefits were also discussed, such as potentially lower house construction costs, with the Master Builders Association arguing that site waste adds 10 per cent onto residential house construction costs. This will benefit first home buyers of new housing as well as government public housing developments.

The qualitative pragmatic research framework was developed next, after a review that also included mixed methods and quantitative research approaches. A model was adapted from the Australian seminal paper by Faniran and Caban

(1998) to determine the potential sources of site waste using a survey questionnaire and data synthesis of model output. This would be carried out using the seminal paper approach but also via thematic analysis. For the front-end site waste minimization strategies research, an abridged USA Yates (2013) model was adopted using a survey questionnaire. The Yates (2013) approach to model synthesis was also adopted as the author's recommended thematic analysis.

Chapter 1, "Introduction", provides the author's reasons for undertaking this research, the development of the initial research question and the development of the option KPIs. Chapter 2 provides the development of the 'long-list' of options and Chap. 3 details the preparation of the 'short-list', as well as "Go Forward" case formulation and the final primary research question. Chapter 4 outlines the author's core study objectives for this research. Chapter 5 sets out the qualitative research methods, including the seminal model approaches and adaption for use for this research survey, and provides the research sub-question on waste sources. These first five chapters mark the mid-point of this research study.

The three-part survey questionnaire developed in Chap. 6, "Results", was circulated via the internet to 102 participants using Qualtrics qualitative survey software. The research "Findings", Chap. 7, provide demographic data from 53 respondents, of whom 51 respondents addressed all questions on sources of waste, down to 48 respondents, who addressed all waste minimization questions. A special feature of this survey was that the participant sample was drawn from a broad spectrum of engineering construction specialists, ranging from top management board directors through to field superintendents, across a whole range of sectors, including engineering design; contractors and builders from commercial, high rise and residential building; heavy civil, mining, process and infrastructure contracting; public and private sector project clients; waste management infrastructure managers; and technical/project management specialists. The results from the potential sources of waste research findings in this book were synthesized, which indicated that the major sources of waste were in alignment with other academic literature.

Design changes were deemed to be a major source of waste, and this book specifically identified client and client-initiated design changes separately. The next highest ranked source of waste, also common with the seminal research, was design and detail errors. Waste from packaging and pallets was the third most commonly identified potential source of construction waste; this aligned with the seminal literature. On the bottom end of the ranking, site waste caused by a criminal activity such as vandalism, and site accidents as a source of site waste, were ranked as negligible by this research and several other academic studies. The front-end waste minimization findings were broadly similar to the seminal research overall, but there were several areas of marked difference that a further academic literature review could possibly validate. Fifty-six per cent of Yates's (2013) USA respondents indicated that their firm was using techniques that improved effective and efficient use of material, labour and equipment resources. Comparatively, for this research only 41.67 per cent of respondents answered in the affirmative.

A comprehensive study by Langston in 2014, which considered relative efficiencies on 337 high-rise building projects in the USA and Australia, found that the

USA projects were considered to have been constructed 37 per cent more efficiently than similar Australian projects. However, this research showed that around 48 per cent of the survey respondents had integrated innovative site waste reduction strategies into their designs and construction processes, while the USA study by Yates (2013) found that only 39 per cent of respondents answered "yes". The global literature review on front-end waste reduction strategies has signalled that a corporate site waste management plan is not considered a priority amongst industry practitioners, but that early design to allow for waste minimization is considered a key strategy; this is reflected in this research.

A key learning from Chap. 8, "Discussion", (regarding the thematic analysis results on the sources of site waste) has been that four of the top six potential sources of waste uncovered in this research were quality control process errors, namely (i) design and detail errors; (ii) procurement ordering and take-off errors; (iii) lack of on-site materials planning and control; and (iv) improper material storage. The major sub-theme, "quality control shortfalls", indicated this common thread running through several sources of site waste categories. The major theme for sources of site waste was "management", and this was also the case for the exhaustive thematic analysis carried out on site waste reduction strategies. In Chap. 8, "Discussion", it is posited that the survey respondents all strongly indicate an emphasis on reducing site waste at the front-end, largely relying on a firm's management as a principal driver of change management. The second most commented upon theme derivation from respondents was the "procurement" theme, this is in line with half of the top six sources of procurement related waste in the study.

Chapter 8, "Discussion", appraisal of the thematic analysis findings for front-end site waste minimization strategies highlighted that the "design" theme was rated second, as per the respondent comments, after the above-mentioned theme on "management". The major sub-theme for this waste minimization analysis was "front-end engineering design" (FEED), which is consistent with Yates' (2013) survey results. Yates (2013) found a strong emphasis by engineering construction practitioners to consider minimizing site waste by using techniques common to the Australian high-rise construction industry, such as precast concrete, and to the heavy construction industry, especially in remote locations, such as off-site modularization and pre-assembly.

The main issue in the "Conclusions" Chap. 9 on waste minimization, points to the requirement for the Federal Government to provide a regulatory framework that establishes a mandatory code of practice for site waste management in Australia and more particularly, for front-end waste. Beneath this overarching practice code, client, design consultant/architect, contractor/builder and vendor/supplier waste management plans would reside within a project waste management plan, tied into the design plan and procurement plan, as part of the project execution strategy. It would be regulatory practice that tenderers submit these waste management plans with their bids. The ultimate responsibility would reside with the client to ensure compliance for site waste minimization. Despite front-end waste minimization strategies having the potential to readily save construction projects two per cent of the contract price, there is a general lack of appetite on behalf of the client, contractor and supplier stakeholders to vigorously pursue these known strategies. This conclusion is supported by several worldwide academic papers, including Australian articles.

The author argues that although there will always be a place for in-vogue downstream recycling/reuse recovery strategies, the collapse in recycling commodity prices, coupled with the recent ban by China on Australian recyclable waste importation, has only decreased the viability of recycling, which is supported by punitive landfill disposal levies by several states. One-third of all solid waste landfill in Australia comes from construction and demolition waste and 40 per cent of all waste is generated by the construction sector. The Australian engineering construction industry has an opportunity to reduce its 13 per cent material wastage rate, which could result in considerable tangible ecological benefits. Carbon emission reduction from manufacturing and transportation; a reduction in both landfill footprint and closure of some of the unregistered/unlined private landfill sites, often favoured by construction contractors; and a reduction in non-productive use of irreplaceable raw material resources, are several positive environmental outcomes that could result from front-end site waste reduction. A two per cent predicted savings in project costs, due to waste minimization of the projected 2017 turnover for all construction in Australia, amounts to a savings of A$4.36 billion per annum. These savings would be enjoyed by both the public/private clients and the contractors/builders. Private residential and public housing prices could fall or stabilize, and clients could reinvest a substantial proportion of these savings into further capital projects, increasing employment of extra construction personnel that would raise GDP, by stimulating private and public sector consumption. The social implications of these front-end waste savings are clear.

A key finding in Chap. 8, "Conclusions", is that strategies are required for innovative solutions for the reduction of packaging and pallet waste. The "innovative solutions" sub-theme ranked third highest commented upon by respondents, which included respondent calls for "packaging and pallet waste" strategies (the third highest ranked sub-theme for the front-end waste minimization thematic analysis and third highest ranked potential source of waste using the Faniran and Caban (1998) sources of waste identification model). In fact, the several respondent comments and their follow up telephone calls (especially for limited footprint inner city high-rise projects) resulted in the author convening a supplementary survey question to gauge whether companies were proactively reducing packaging and pallet waste; 12 of the 17 respondents, from a sample of 50 participants, answered "no". The textbook concluded that a small cohort of professional, contractor/builder, supplier, peak body, academic and government panellists should establish a working group to address this urgent issue requiring a solution. This strategy is in line with the Federal Government's April 2018 announcement that it intends to tackle the generic issue of packaging waste in light of the current commercial pressures on the Australian recycling business.

The final chapter of this textbook is Chap. 10, "Recommendations". A major recommendation of this study considers possible areas of further academic research to advance the knowledge of construction site waste minimization. Possible future topics for further waste management research could be: (i) a detailed cost–benefit analysis; (ii) a case study applying the learning from this research, preferably of a high-rise building, which would allow close review of waste inventories and the opportunity to apply several of the proposed front-end waste reduction techniques outlined herein; and (iii) dynamic modelling and/or optimization modelling of site waste. Another significant recommendation considers the efficacy of government generated waste management data. The 2016 *"State of the Environment Overview"*, tabled in March 2017 to Parliament in Canberra, contained data going back as far as 2010/2011 with different data dates for different states. The literature has shown that Australian academics, consulting professionals and waste management professionals, are very sceptical about alleged government manipulation of waste data to suit their political stance. All nine Australian governments have a duty of care to compile a set of data statistics annually so that public and private stakeholders can make strategic value decisions with the necessary confidence in the quality of the data.

References

ABS. (2015). *Annual report, 2014–2015,* cat. 1000.1, Canberra: ABS.

Austrade. (2018). *Building & construction capability report.* https://www.austrade.gov.au/International/Buy/Australian-industry-capabilities/Building-and-Construction, viewed September 6, 2018.

Faniran, O. O., & Caban, G. (1998). Minimizing waste on project construction sites. *Engineering Construction and Architectural Management, 17*(1), 57–72.

Gruszka, A. (2017). *How technology is transforming Australia's construction sector* (StartupAUS Report) pp. 1–67. http://www.apcc.gov.au/ALLAPCC/SAUS_ConTech_Report_2017.pdf, viewed September 6, 2018.

Kepner, C. H., & Tregoe, B. B. (2013). *The new rational manager.* Princeton, USA: Kepner Tregoe Inc. Publishing.

Yates, J. (2013). Sustainable methods for waste minimisation in construction. *Construction Innovation, 13*(3), 281–301.

Appendix B
Kepner Tregoe Decision Analysis 'Short-List' Scorecard—Options A, B, C and D

© Springer Nature Switzerland AG 2019
P. G. Rundle et al., *Effective Front-End Strategies to Reduce Waste on Construction Projects*, https://doi.org/10.1007/978-3-030-12399-4

Option No. "A" Optimal Site Hours	"A" Marking Criterion Weighting	Ken Doust	Ali Reza	Kevin Bradley	Gary Murphy	D. Mulvihill	Allan Morgan	Graham Stacey	Jarrod Hayton	"B" Peer Panel Member Option Score	A x B Option Criterion Total
Positive impact of option on efficiency defined as: meeting all internal requirements for cost, margins, asset utilization and other related efficiency measures, in Australian construction industry	5	5	5	5	5	5	5	5	5	5	25
Positive impact of option on effectiveness defined as : satisfying or exceeding customer needs with a compliant product, in Australian construction industry	5	5	5	5	5	5	5	5	5	5	25
Social and environmental benefit of option	4	5	5	3	3	4	4	4	4	4	16
Ease of option application and roll-out in the construction industry	3	4	4	4	4	4	3.5	3.5	4	3.875	11.625
Ease of option application and roll-out in the construction industry	3	3	3	2	4	4	3	3	3	3.125	9.375
Void in the literature	5									5	25
											Total = 112

Option No. "B" Site Wastage	"A" Marking Criterion Weighting	Ken Doust	Ali Reza	Kevin Bradley	Gary Murphy	Damien Mu	Allan Morgan	Graham Stacey	Jarrod Hayton	"B" Peer Panel Member Option Score	A x B Criterion Total
Positive impact of option on efficiency defined as: meeting all internal requirements for cost, margins, asset utilization and other related efficiency measures, in Australian construction industry	5	5	5	5	5	5	5	4	5	4.875	24.375
Positive impact of option on effectiveness defined as : satisfying or exceeding customer needs with a compliant product, in Australian construction industry	5	5	5	5	5	5	5	4	5	4.875	24.375
Commercial benefit of option to construction industry stakeholders including contractors, suppliers, consultants and customers	4	5	4	5	4	4	5	4	5	4.5	18
Social and environmental benefit of option	3	5	3	5	4	5	4	3	5	4.25	12.75
Ease of option application and roll-out in the construction industry	4	4	3	3	4	3	3	3	2	3.125	12.5
Void in the literature	5									5	25
											Total = 117

Option No. "C" Site Safety Regulation Rationalization	"A" Marking Criterion Weighting	Ken Doust	Ali Reza	Kevin Bradley	Gary Murphy	Damien Mu	Allan Morgan	Graham Stacey	Jarrod Hayton	"B" Peer Panel Member Option Score	A x B Option Criterion Total
Positive impact of option on efficiency defined as: meeting all internal requirements for cost, margins, asset utilization and other related efficiency measures, in Australian construction industry	5	3	4	4	3	3	3	4	5	3.625	18.125
Positive impact of option on effectiveness defined as : satisfying or exceeding customer needs with a compliant product, in Australian construction industry	5	3	4	5	3	3	3	4	5	3.75	18.75
Commercial benefit of option to construction industry stakeholders including contractors, suppliers, consultants and customers	4	3	4	4	3	4	3	4	4	3.625	14.5
Social and environmental benefit of option	3	3	4	4	5	4	3	4	3	3.625	10.875
Ease of option application and roll-out in the construction industry	2	3	1	3	3	2	2	1	2	2.125	4.25
Void in the literature	5									3	15
											Total = 82

Option No. "D" Sustainable Labour Construction Force	"A" Marking Criterion Weighting	Ken Doust	Ali Reza	Kevin Bradley	Gary Murphy	Damien Mu	Allan Morgan	Graham Stacey	Jarrod Hayton	"B" Peer Panel Member Option Score	A x B p_ on Criterion Total
Positive impact of option on efficiency defined as: meeting all internal requirements for cost, margins, asset utilization and other related efficiency measures, in Australian construction industry	5	5	5	5	3	4	4	5	5	4.5	22.5
Positive impact of option on effectiveness defined as : satisfying or exceeding customer needs with a compliant product, in Australian construction industry	5	5	4	5	3	4	4	5	5	4.375	21.875
Commercial benefit of option to construction industry stakeholders including contractors, suppliers, consultants and customers	4	5	3	3	3	4	3	5	3	3.625	14.5
Social and environmental benefit of option	3	4	3	4	3	4	3	4	3	3.5	10.5
Ease of option application and roll-out in the construction industry	2	2	3	3	3	2	2	3	3	2.625	5.25
Void in the literature	5									3	15
											Total = 90

By J. Novak, August 2017

Appendix C: Conclusions: Research Methodology and Void in the Literature

C.1. Survey Saturation and Sample Size Conclusions

Pursuant to Sect. 6.3, "Survey Saturation", of this textbook, the author has adopted an empirical approach to determining saturation. For this research, a sample size of 53 respondents with a minimum of 48 completed surveys exceeds the average qualitative methods requirement of 25 completed respondent surveys (Charmaz 2006; Creswell 1998; Mason 2010). Notwithstanding the empirical approach to saturation sampling quantum adopted in this research by the author after duly interpreting a rich pool of respondent survey data (refer to Chap. 6, "Results") via thematic analysis in Chap. 7, "Findings", of this book, the theoretical saturation approach to classic qualitative research has also been complied with in this research. Morse (1995, p. 147) defines saturation as having "data adequacy and operationalized as collecting data until no new information is obtained", which is in alignment with the seminal saturation approach to sampling as outlined by Glaser and Strauss (1967). Refer to Sect. 8.1.1, "Survey Saturation Discussion", for the full details of this analysis.

Refer to Sect. 8.1.2, "Survey Sample Size Discussion", of this textbook, which details the reasons why a sample size of around 100 survey questionnaires distributed to participants for a respondent rate of 25 per cent was applicable in order to obtain saturation, and for the provision of a pool of adequate data for thematic analysis. It has also been concluded that a proactive approach to the management of survey sampling is necessary to guarantee a sufficient size of the sample by tailoring covering letters to each participant; developing a strategy for sample participants during the early phases of a research project; and managing the survey process via such actions as following up on respondent questions by telephone. As an empirical check of the appropriateness of a 100-person sample pool, Coakes and Ong (2011) nominated a sample size of 100 or more being appropriate, albeit for a human factors approach to waste management strategy review, as per the Udawatta et al. (2015) construction waste management study.

© Springer Nature Switzerland AG 2019
P. G. Rundle et al., *Effective Front-End Strategies to Reduce Waste
on Construction Projects*, https://doi.org/10.1007/978-3-030-12399-4

C.2. Demographics Conclusion

The benefit of using a stratified sampling strategy divided into homogenous sub-groups has not only reduced a propensity for sampling bias but has also positively benefited this research because the respondents have provided a broad spectrum of opinions from their own fields of the engineering construction industry. This has provided significant insight into the broadly agreed sources of site waste and site waste mitigation strategies. Forty-six per cent of the baseline Yates (2013) survey participants and 46 per cent of the respondents from this research on the topic of site waste minimization strategies agree on their merit. The top three (most likely) sources of site waste and the bottom two (least likely) sources of site waste for respondents polled in the Faniran and Caban (1998) survey on sources of site waste concurs with the survey results for this textbook.

This research has shown that there is significant agreement throughout the Australian engineering construction industry ranging from the company board level to project managers, site superintendents, engineering design managers, major capital works public and private sector clients, project management professionals and landfill/recycling plant operators, as to the major sources of waste, as well as mitigation strategies for site waste reduction, that can be benchmarked against previous academic papers and thus triangulated.

This broad spectrum demographic approach has harvested much topical insight, which provided a rich harvest of data for use, particularly in the thematic analysis. This demographic spectrum approach to sample conformation has not been used as extensively by any other academic papers on site waste evaluated in this research. However, it is strongly recommended that it be utilized for future research in this field of study.

C.3. Pragmatic Qualitative Research Framework Conclusion

Refer to Sect. 8.4, "Pragmatic Qualitative Research Framework Discussion", for the full details of Table A, "Summary of Research Outcomes of the Framework for a Pragmatic Qualitative Research Approach".

Table A, "Summary of Research Outcomes of the Framework for a Pragmatic Qualitative Research Approach", adapted from Pope et al. (2000) from British research standards, clearly concludes that the research framework has been stringently followed and has consequently yielded rigorous results using the academic research methodologies nominated above. This pragmatic qualitative framework research method is suitable for further academic studies on construction site waste, using thematic analysis to interpret industry professional and practitioner participant survey and interview data. This approach has proven to be a valuable qualitative research method to collate and identify key themes and factors from a rich pool of respondent data. The thematic approach would be a valuable research tool to study other similar engineering science academic studies.

Table C.1 Summary of research outcomes of the framework for a pragmatic qualitative research approach

Standards	Terms associated with qualitative research	Research outcomes
Framing	–	Refer to Sect. 8.4.2, "Framing Discussion", which provides details on what is known and what this study builds on, along with the information never described before
Reliability	Procedural trustworthiness	Refer to Sects. 8.1 and 8.2, which justify the size and nature of the sample Refer to Sect. 8.3, which addresses the research question
Validity	Trustworthiness of the findings	Refer to Sect. 8.4.3, "Validity Discussion", which details how triangulation of data from academic literature and previous academic research was used to correlate this research data. This section also details how sample processes have demonstrated catalytic validity
Generalisability	Transferability	Refer to Sect. 8.4.4, "Generalizability Discussion", which details exactly how seminal international models adopted for this research have allowed for transferability. This section also details how KPIs of project success are the practical application of this research for use in the field for practitioners and academics
Objectivity	Permeability/trustworthiness of researcher	Refer Sect. 8.4.5, "Objectivity Discussion", for full details. During 40 years as a civil engineer, the author has worked on over 20 feasibility studies and this has taught him not to assume a result but to allow the study to provide the options. A reflexivity journal was kept, and details are contained in Tables 7.7 and 7.40 of this textbook with the author's key reflections on the respondent comments covering the questionnaire on sources of site waste and the questionnaire on site waste minimization strategies. The author has also taken self-awareness training as part of a Masters of Business Administration and has used these practices for reflecting on this research approaches. An independent facilitator and two independent peer review panels containing eminent specialists were adopted in this research. Research findings were rigorously tested against pre-existing academic research, with close correlations determined

By: P. Rundle, 2018. Adapted from Pope et al. (2000)

C.4. Void in the Australian Literature

With the publication of this textbook, another fragment of the void in the Australian academic literature on physical construction waste shall be filled.

Refer to Sect. 8.3, "Research Question Discussion", which provides an explanation as to why this textbook has met the research question and the sub-research question criteria.

The reader's attention is drawn to Sect. 5.1.1, "Site Waste Sources and Minimization", in which both the differences and the complementary nature of previous Australian academic literature on construction site waste compared to this textbook are discussed. This research has added to the seminal work of Faniran and Caban (1998) by updating their research 20 years on and by making recommendations of minor revisions to their sources of waste identification model. This research has shown that the sources of waste in Australia in 2018 are very similar to those in 1998.

Sugiharto et al. (2002) research was a valuable paper for the correlation of data with this textbook. Udawatta et al. (2015) pointed the way in their research conclusions for further research, which regarded using a stratified participant survey sample that was adopted for this research. The qualitative thematic analysis used for data synthesis in this research, compared to the statistical factor analysis used by Udawatta et al. (2015), yielded a similar primary conclusion on site waste management; that is, the need for greater regulatory involvement and the importance of a construction waste management plan across key project stakeholders.

This study has shown the value of carrying out a research scope that considers both the sources of site waste and the suitable strategies for front-end waste minimization. It also demonstrates how the use of thematic analysis assists in uncovering the inter-relationships between construction waste sources and a strategy to remove, or minimize, that source of waste.

It is hoped that higher research by a future candidate, of at least one of the suggested topics noted in Chap. 10, "Recommendations", is undertaken to move the study of construction waste forward and to add to the knowledge acquisition database contributions of the above noted research of Yates (2013) and of many others.

References

Charmaz, K. (2006). *Constructing grounded theory: A practical guide through qualitative analysis*. Thousand Oaks, USA: Sage.

Coakes, S. J., & Ong, C. (2011). *SPSS: Analysis without anguish version 18.0 for Windows*. Milton, Queensland: Wiley.

Creswell, J. W. (1998). *Qualitative inquiry and research design: Choosing among five traditions*. Thousand Oaks, USA: Sage.

Faniran, O. O., & Caban, G. (1998). Minimizing waste on project construction sites. *Engineering Construction and Architectural Management, 17*(1), 57–72.

Glaser, B., & Strauss, A. (1967). *The discovery of grounded theory.* Hawthorne, NY: Aldine Publishing Company.

Mason, M. (2010). Sample size and saturation in PhD studies using qualitative interviews. *Forum for Qualitative Social Research, 11*(3), 1–19.

Morse, J. M. (1995). The significance of saturation. *Journal of Qualitative Health, 5*(2), 147–149.

Pope, C., Ziebland, S., & Mays, N. (2000). Analysing qualitative data. *British Medical Journal, 320,* 114–116.

Sugiharto, A., Hampson, K., & Sherid, M. (2002). Non value adding activities in Australian construction projects. In: *Paper presented to the International Conference for Advancement in Design, Construction Management and Maintenance of Building Products*, Griffith University, Australia, 2002.

Udawatta, A., Zuo, J., Chiveralls, K., & Zillante, G. (2015). Improving waste management in construction projects: An Australian study. *Resources, Conservation and Recycling, 101,* 73–83.

Yates, J. (2013). Sustainable methods for waste minimisation in construction. *Construction Innovation, 13*(3), 281–301.

Bibliography

Bamford, D. R., & Forester, P. (2013). Managing planned and emergent change within an operations management environment. *International Journal of Operations & Production Management, 23*(5), 546–564.

Bossink, B. A. G. (2004). Managing drivers of innovation in construction networks. *Journal of Construction Engineering and Management, 130*(3), 337–343.

Bullock, R., & Batten, D. (1985, December). It's just a phase we're going through: A review and synthesis of OD phase analysis. *Group and Organizational Studies, 10*, 383–412.

Burnes, B. (2004). Kurt Lewin and the planned approach to change: A re-appraisal. *Journal of Management Studies, 41*(6), 977–1002.

Burnes, B. (Ed.). (2004). *Managing change: A strategic approach to organizational dynamics.* Harlow, England: Pearson Education Limited.

By, R. (2005). Organizational change management: A critical review. *Journal of Change Management, 5*(4), 369–380.

Cameron, R. (2015). *Mixed methods research workshop, Australia New Zealand Academy of management conference*, July 2. Paper presented to Curtin University School of Business, South Australia.

Chau, D. (2017). Construction data hints at GDP slowdown: ABS. In *ABC News*, May 25. http://www.abc.net.au/news/2017-05-25/construction-work-done-gdp-impact/8559176, viewed May, 2017.

CII. (2017). *Home page.* https://www.construction-institute.org, viewed January, 2018.

CIPD. (2017). *PESTLE analysis.* https://www.cipd.co.uk/knowledge/strategy/organisational-development/pestle-analysis-factsheet, viewed May, 2017.

Coghlan, D., & Brannick, T. (2003). Kurt Lewin, the practical theorist for the 21st century. *Irish Journal of Management, 24*(2), 31–37.

Cooper, R. G., Edgett, G. C., & Kleinschmidt, E. J. (2012). New problems, new solutions: Making portfolio management more. In *Industrial Research Institute*.

Core Logic. (2016). *Cordell cost guide.* https://www.corelogic.com.au/products/cordellcostguides, viewed October–December, 2017.

Davis, P. (2017). War on waste: The battle to minimize waste on NSW construction projects. In *Australian Institute of Builders*, no. 4.

Dunphy, D., & Stace, D. (1993). The strategic management of corporate change. *Human Relations, 46*(8), 905–918.

Emerson, R. M. (1976). Social exchange theory. *Annual Review of Sociology, 2*, 335–362.

Fellows, R. T., Langford, D., Newcomb, R., & Urry, S. (2009). *Construction management in practice.* UK: Blackwell Science.

© Springer Nature Switzerland AG 2019

P. G. Rundle et al., *Effective Front-End Strategies to Reduce Waste on Construction Projects*, https://doi.org/10.1007/978-3-030-12399-4

Formoso, C. T., Isatto, E. L., Hirota, E. (1999, July). *Method for waste control in the building industry*. Paper presented to 7th Annual Conference of the International Group for Lean Construction, University of California, Berkley.

Hiete, M., Stengel, J., Ludwig, J., & Schultmann, F. (2011). Matching construction and demolition waste supply to recycling demand: A regional management chain model. *Building Research & Information, 39*(4), 333–351.

Ibrahim, M. (2016). Estimating the sustainability returns of recycling construction waste from building projects. *Sustainability Cities & Society, 23*, 78–93.

Kester, L., Hultink, E., & Griffin, A. (2013, June). Empirical exploration of the antecedents and outcomes of NPD portfolio success. In *International Product Development Management Conference, Paris, France*.

Leech, N., & Onwuegbuzie, A. (2008). A typology of mixed methods research designs. *Quality and Quantity, 43*(2), 265–275.

Love, P. (2006). Melbourne celebrates the 150th anniversary of its eight hour day. *Labour History, 91*.

MacDonald, C., Walker, D., & Moussa, N. (2013). Towards a project alliance value for money framework. *Facilities, 31*(5/6), 279–309.

Master Builders Association of Victoria (MBAV). (2004). *The resource efficient builder—A simple guide to reducing waste*. MBAV Guideline Publication.

McShane, S., Olekalns, M., & Travaglione, T. (Eds.). (2013). *Organisational behaviour: Emerging knowledge, global insights*. Australia: McGraw Hill.

Oliver, R. (1974). Expectancy is the probability that the individual assigns to work effort being followed by a given level of achieved task performance—Expectancy theory predictions of salesmen's performance. *Journal of Marketing Research, 11*, 243–253.

Peteraf, M. (1993). The cornerstones of competitive advantage: A resource based view. *Strategic Management Journal, 14*(3), 179–191.

Project Management Body of Knowledge (PMBOK). (2000). A guide to project management body of knowledge. *Journal of Project Management Institute*.

Polkinghorne, D. (2011). Phenomenological research methods. In R. Vale & R. Halling (Eds.), *Existential-phenomenological perspectives in psychology* (pp. 41–46). New York, USA: Plenum Press.

Regulation Taskforce. (2006). *Rethinking regulation: Report of the taskforce on reducing regulatory burdens on business* (Report to the Prime Minister and the Treasurer, Canberra) January, http://www.pc.gov.au/research/supporting/regulation-taskforce/report/regulation-taskforce2.pdf, viewed May, 2017.

Sullivan, G., Barthorpe, S., & Robbins, S. (2011). *Managing construction logistics*. USA: Wiley.

Van Eerde, W., & Thierry, H. (1966). Vroom's expectancy models and work related criteria: A meta-analysis. *Applied Psychology, 81*, 575–586.

Wahi, V., Joseph, C., Tawii, R. Z., & Ikan, R. (2016). Critical review on construction waste control practices: Legislative and waste management perspective. *Procedia—Social and Behavioural Sciences, 224*, 276–283.

Walker, A. (Ed.). (2011). *Organizational behaviour in construction*. USA: Wiley.

Welsh, M., Al Hakim, A., Ball, F., Dunstan, J., & Begg, S. (2011). Do personality traits affect decision making ability: Can MBTI type predict biases. *APPEA Journal, 51*(1), 359–368.

Index

© Springer Nature Switzerland AG 2019
P. G. Rundle et al., *Effective Front-End Strategies to Reduce Waste on Construction Projects*, https://doi.org/10.1007/978-3-030-12399-4

Printed in the United States
By Bookmasters